LAKE BOGA AT WAR

THE INSIDE STORY OF THE SECRET RAAF INLAND FLYING BOAT BASE IN WWII

REVISED EDITION

BRETT FREEMAN

AVONMORE BOOKS

Lake Boga at War

The inside story of the secret RAAF inland flying boat base in WWII

Revised Edition

Brett Freeman

ISBN: 978-0-6457004-2-8

Published 2023 by Avonmore Books

Avonmore Books
PO Box 217
Kent Town
South Australia 5071
Australia

Phone: (61 8) 8431 9780
avonmorebooks.com.au

 A catalogue record for this book is available from the National Library of Australia

Cover design & layout by Diane Bricknell

© 2023 Avonmore Books

No part of this book may be reproduced or transmitted in any form or by any means, electronic or mechanical, including photocopying or recording, or by any information storage and retrieval system, without permission in writing from the publisher.

Front cover artwork:

Aviation artist Brian Evans vividly depicts No. 1 Flying Boat Repair Depot on 23 December 1943. Martin Mariner flying boats have just arrived from the US as have three of the five Catalinas that alighted at Lake Boga on that day. While a Dornier is taxiing, Flight Lieutenant Mike Seymour DFC is at the controls of an early RAAF Catalina, A24-4.

Back cover photo:

PBY-5A A24-69 of No. 11 Squadron being serviced at Lake Boga in August 1944. The photo gives a good view of the open hangars which were designed so that maximum light was obtained for intricate repairs. (AWM)

Contents

Foreword to Revised Edition	5
Abbreviations	6
Author's Introduction to 1995 Edition	7
About the Author	8
Chapter 1 – 1942	9
Chapter 2 – 1943	29
Chapter 3 – 1944	47
Chapter 4 – 1945	63
Chapter 5 – 1946-1947	73
Index of Names	77

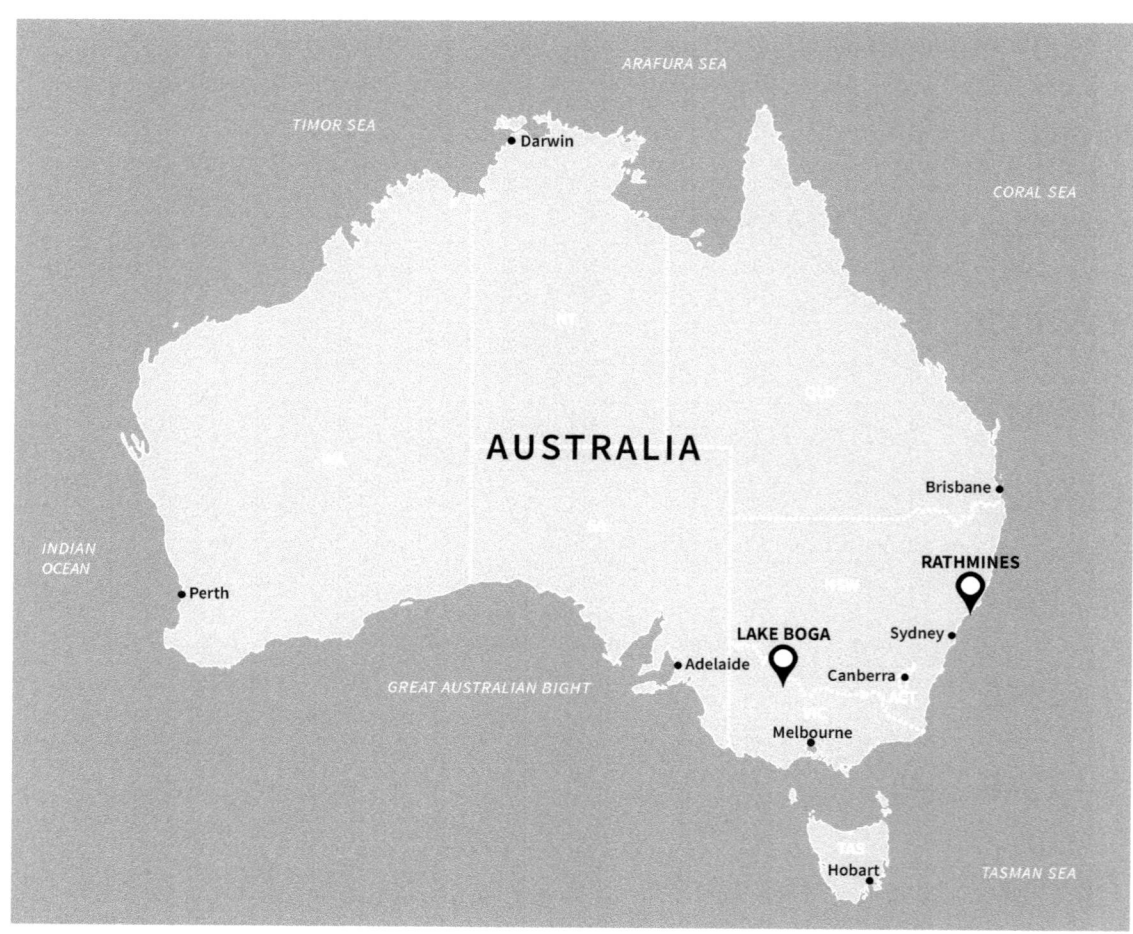

During WWII, flying boats played a key role in keeping watch over the waters to the north of Australia and the islands beyond. Often these areas had few if any airfields, especially during the early part of the Pacific War. The great range of the flying boats also enabled them to conduct offensive missions over enemy territory, mainly at night. Later minelaying missions were flown, and flying boats also took on an important search and rescue role. At the start of the war the key RAAF flying boat base was at Rathmines on the New South Wales coast. However, in 1942 such a location lay exposed should Japanese forces move south, and a more secure inland location at Lake Boga in Victoria was chosen that became home to No. 1 Flying Boat Repair Depot.

Foreword to Revised Edition

The original edition of *Lake Boga at War* was published in 1995 by the author, the late Brett Freeman. It was a remarkable book that met a massive public appetite for a wartime history of an otherwise obscure inland flying boat base that was classified "top secret" for most of the war years. The book was reprinted in 2017 and in total has sold 5,000 copies – a wonderful result for a first-time author of a self-published book. *Lake Boga at War* is surely one of the most successful Australian aviation history books of its era.

Such a work can only be forthcoming from an author with a lifetime of passion and research in respect to the subject at hand. Brett was an eleven-year-old local boy who witnessed the establishment of the Lake Boga base in 1942 and then all of the subsequent activity until its disbandment in 1947. In the decades since Brett ran a business in Swan Hill and was hence able to know personally many dozens of those that had worked at the base during the war or who had visited it as aircrews. The latter comprised members of the RAAF as well as Dutchmen and Americans from both the USN and the USAAF.

Brett's work was based on the Operations Record Books of No. 1 Flying Boat Repair Depot (No. 1 FBRD) which in the pre-internet days were only accessible by visiting archives in Canberra. However, the beauty of his work was the matching of these detailed records with a vast amount of oral history gleaned from his multitude of contacts. With the passing of time, such an approach is impossible today.

The original edition of *Lake Boga at War* was a 296-page hardback, and today it is uneconomic to republish the book in that form. This revised edition is somewhat smaller, to enable it to reach a new generation of readers. The approach has been to retain Lake Boga-specific material wherever possible. The biggest changes are the omission of the original Chapter 1 and the Epilogue. The original Chapter 1 covered the period 1939-1941, prior to the establishment of No. 1 FBRD in 1942. The Epilogue covers activities after 1947, mainly pertaining to the post-war fates of various aircraft that had passed through the base and associated personnel. In addition, some commentary on the general war situation and the wider biographical details of certain personnel has been omitted. Some additional photographs, not available to the author in the 1990s, have been added.

Sir Richard Kingsland wrote a forward to the 1995 edition which ended with the following lines which remain prescient today:

> How easily this story could have been lost. Allied flying boat air and ground staff owe a great debt to Brett Freeman for his labours.

This new edition has been produced with the kind assistance and cooperation of Sally Fearn (Brett's daughter), Lois Freeman (Brett's wife) and the Lake Boga Flying Boat Museum. I trust it preserves the best of the original edition and serves to stimulate ongoing interest in a fascinating yet little-known part of Australia's wartime history.

Peter Ingman - editor
Avonmore Books
April 2023

Abbreviations

AC1	Aircraftman, Class 1
ACW	Aircraftwoman
AEA	Air Efficiency Award
AFC	Air Force Cross
AIF	Australian Imperial Force
ASO	Assistant Section Officer (WAAAF)
ASR	Air Sea Rescue
AWM	Australian War Memorial
CCC	Commonwealth Construction Corporation
CMU	Care and Maintenance Unit
CO	Commanding Officer
DCA	Department of Civil Aviation
DFM	Distinguished Flying Medal
DOA	Department of Air
DOI	Department of the Interior
DSO	Distinguished Service Order
DWB	Directorate of Works and Buildings
FBMU	Flying Boat Maintenance Unit
FBRD	Flying Boat Repair Depot
HP	Horse Power
HQ	Headquarters
LAC	Leading Aircraftman
Lieutenant (jg)	Lieutenant (junior grade)
MID	Mentioned in Dispatches
MO	Medical Officer
NEI	Netherlands East Indies
OTU	Operational Training Unit
POW	Prisoner of War
RAAF	Royal Australian Air Force
RSL	Returned Service League
SP	Service Police
SRWSC	State Rivers and Water Supply Commission
US	United States
USAAF	United States Army Air Force
USN	United States Navy
VC	Victoria Cross
WAAAF	Women's Auxiliary Australian Air Force
WAG	Wireless Air Gunner
W/T	Wireless Telegraphist

Author's Introduction to 1995 Edition[1]

Driving along the Murray Valley Highway, its route paralleling the course of Australia's greatest river, you pass through the small north-western Victorian township of Lake Boga, ten miles from Swan Hill.

To the north of the town, a high rusting barbed wire fence still stands, surrounding an area of stunted Yanga salt bush among which a windowless concrete blockhouse sits silently with its faded shades of khaki camouflage.[2]

'What was there, Dad?' asks an enquiring child, as a family drives past. 'Oh er … they er … used to fix flying boats during the war' is often the reply.

Quite correct. RAAF personnel did repair and maintain Allied flying boats at this site during WWII, for this is the site of No. 1 Flying Boat Repair Depot (No. 1 FBRD), Lake Boga.

I have always had an interest in the tales of 'the Air Force at Boga during the war'. In March 1942, as an eleven-year-old, I witnessed the arrival of the first flying boat to alight on Lake Boga, and in May of that year was intrigued by a double page diagram of a Catalina flying boat published in *Flight* magazine. As I pasted the diagram in my aircraft scrap book, I was not to know just how many Catalinas and other flying boats would fly into the district during WWII. In 1944, I was allowed to inspect this still secret facility and in 1946, together with young friends, I crawled through the stilled, stored aircraft to my heart's content.

During the war years, several RAAF personnel became family friends and while working on this volume I managed to trace the NEI Catalina flight engineer who taught me to count in Dutch during his 1943 stopovers.

Five fulfilling years have now passed researching many aspects surrounding No. 1 FBRD which I have chronicled as accurately and comprehensively as possible.

At the time of the Japanese entry into WWII in December 1941, Lake Boga and the adjacent town of Swan Hill had populations of 220 and 4,800 respectively. The region known as the Mallee was and remains primarily, a grain and wool producing area, with its irrigation supporting dairying, animal breeding, and vegetable and fruit production, much of the latter having been developed by solider settlers after WWI.

These two towns, among others servicing this rural region, suddenly found themselves host to nigh on 1,000 RAAF and WAAAF personnel, plus visiting aircrew from the RAAF, USN, USAAF and the Dutch naval air service.

The central thread, around which this story is woven, is based on information extracted from the Operations Record Books of No. 1 FBRD during 1942-1947. These records are those that formed the depot's daily diary, signed by one temporary and eight permanent commanding officers.

For the sake of this narrative, it is fortunate that in most instances the life of an aircraft is more completely documented than that of a human being. Aeroplanes have a birth and death certificate, plus lifelong certificates of airworthiness. These particulars, when married to information extracted from aircrew flight logs, produce a vivid image of what actually occurred, where it occurred, and when.

This then is a story backgrounded by contemporary worldwide events at a time when many farmers still used horses as a primary power and transport source. It is an account of the life of No. 1 FBRD, its personnel, its visiting aircrew and aircraft, and its adjacent communities: an account of Australia, Swan Hill and … Lake Boga at War!

Brett SW Freeman
Swan Hill

---

1 This is an edited version of the original.
2 Today the location is the home of the Lake Boga Flying Boat Museum, which opened in 2012. Among the exhibits is the beautifully restored Catalina A24-30.

About the Author

Born in Swan Hill, Brett Freeman was educated at Swan Hill primary and secondary school as well as Melbourne Grammar School. He spent his working life in the retail fashion business.

Community minded, the author has been a member of several service organisations and is a past president of the Swan Hill Rotary Club. Having an interest in history, Brett Freeman served for 25 years on the board of Australia's first pioneer village, Swan Hill Pioneer Settlement. Other interests include travel and snow skiing.

CHAPTER 1
1942

Early 1942 saw Australia under the threat of Japanese invasion, and various plans were being made for the defence of the continent. With RAAF seaplane operations centred at Rathmines on the New South Wales coast, the Air Board was intent on the establishment of a new seaplane maintenance depot and safe haven for its RAAF flying boats.

The existence of Lake Boga as a potential site for flying boat activity had been known to the government as early as 1938. At that time various approaches, including that of the Lake Boga Progress Association, were made in support of a flying boat training base and/or air mail terminal. "Representations have already been made to the local Parliamentary Members, and to the Prime Minister and Sir Earle Page, and they have all offered their support," advised a July 1938 article in the *Swan Hill Guardian*.

In early March 1942 a Department of Civil Aviation (DCA) inspection party arrived in Swan Hill where they met with State Rivers and Water Supply Commission District Officer Jack D Wallis. They made an inspection of both Lake Boga and Kangaroo Lake, with a preference for Lake Boga. Originally natural lakes, both bodies of water were now fed from the Murray River through the Torrumbarry irrigation system.

Lake Boga possessed adequate vacant land along the lake's foreshore, an adjacent rail head and highway, electric power from the Swan Hill power station, and a large body of obstacle free water surrounded by flat countryside, all the ingredients necessary for the safe alighting and service of flying boats.

While examining an alternate body of water in the Waranga Basin, a member of the DCA party wrote to Wallis from Tatura 9 March, concluding his letter:

> In case of sudden emergency (which might arrive at any time) I should be obliged if you could get one of the Commission's staff to go over the lake and remove any stakes that may have been put in by fishermen or the sailing people. It may be necessary to land one of the big flying boats on the lake, even within a day or two.

On 13 March, Wallis received a telegram from Sydney asking him to arrange for the arrival at Lake Boga of an amphibian aircraft carrying Qantas personnel.

With the assistance of Messrs Fred Petzke and Roy Irvine, the lake was cleared of obstructions including fishing net stakes and the Lake Boga Yacht Club's pile mounted judging box. Petzke was a successful Swan Hill plumber and tank maker, Irvine the local Ford agent who promoted business by nailing old T model Ford doors to roadside trees throughout the region, inscribed, "FORD - Phone Swan Hill 123".

The following Sunday afternoon a strange biplane flying boat, with a large back to front motor (an RAAF Walrus), circled, then alighted upon Lake Boga, taxiing towards the recreational reserve. A crowd had gathered by the time Flight Lieutenant John McMahon cut the spluttering motor and anchored the Supermarine Walrus amphibian

On 15 March 1942, Hudson Fysh and Lester Brain from Qantas alighted on Lake Boga in a Walrus amphibian. A later photo. (Gordon Myers)

in shallow water. The Qantas General Manager, Hudson Fysh, together with the Operations Manager, Lester Brain, then emerged from the craft and came ashore via a boat where they were welcomed by Wallis.

A short time later, when the Qantas party's inspection of the Lake Boga sites was complete, the Walrus took off for the Swan Hill aerodrome. Upon landing, the quaint amphibian taxied to the tin shed terminal. At the conclusion of a meeting at the State Rivers office that Sunday evening, Lake Boga was further firming as the favoured site for the DCA base.

With the surrender of the Netherlands East Indies and the landing of Japanese troops in mainland New Guinea in March, the war situation was becoming critical. Black outs had now been imposed on many Australian cities and towns, railway station names were suddenly removed, prefabricated air raid shelters were advertised in newspapers as were shielded black-out hoods for motor vehicle headlights. In many areas of Australia air raid trenches were dug.

Jack Wallis, SRWC Regional Engineer, Swan Hill. (Peter Wallis)

By mid-March the Commonwealth Department of the Interior (DOI) funded and issued instructions for the laying down of a provisional flying boat slipway and installation of flying boat mooring buoys at Lake Boga. Civil engineer Vladimir Michels of the DOI Works and Services Branch arrived at Lake Boga on 26 March. Ten days earlier, the Department had appointed Michels responsible for the installation, supervision and maintenance of all seaplane moorings in Victoria and Tasmania.

Together with his Russian parents, Michels had migrated as "White" refugees from Bolshevik Russia, moving initially to China, then to Australia in 1926. By 1938 he had completed his degree course in Civil Engineering with honours at Melbourne University. His tasks at Lake Boga, according to his works diary, were "urgent and to proceed like wildfire", particularly in respect to the temporary slipway.

Vladimir Michels, SRWSC Engineer. (Vladimir Michels)

With quiet efficiency, Swan Hill Shire and acting borough engineer, Cyril Lowenstern, provided Michels with every assistance, assembling necessary equipment and providing him with heavy concrete sinkers as buoy anchors and transporting them to the site. Second hand chains, swivels and shackles then arrived from Ma Dally's noted North Melbourne junk yard. Other scarce or substitute items were hunted down in Swan Hill.

The Melbourne Harbour Trust lent diving equipment together with an experienced diver. By midday on 28 March the mooring systems had been set out on the foreshore ready for installation. Offshore winds presented difficulties in laying, making the Qantas launch, which arrived two days later, indispensable.

March 31 saw the special Qantas mooring buoys, with their protective fenders, successfully anchored in position. The chain mesh matting having duly arrived, construction of the temporary slipway proceeded apace, Michels supervising the workmen who assisted the diver. All worked late. The diver drove large steel pegs into the lakebed, thus anchoring the edges of the tensioned chain mesh matting. Next morning Michels, "tired and with chapped hands", departed by train to Melbourne.

By April, the intention to establish an RAAF repair depot and training facility additional to Lake Boga's DCA facility had been confirmed and consultation with Qantas in overall planning continued. In early 1942 Qantas had lost five flying boats to enemy action or war service. At the behest of Hudson Fysh, the Director General of Civil Aviation, AB Corbert, had been quick to fund the establishment of an emergency mooring, minor maintenance, and refuelling facility at Lake Boga "in case Rose Bay becomes untenable". While cautioning Qantas regarding the economics of establishing a large flying boat facility so far inland, Corbert advised the DCA on 24 March:

> If the Air Board requires an alternative establishment to service flying boats, "Catalinas" or "Seagulls" [Walrus] there may be a case for building such a base at Lake Boga; but again it must be considered if such a place can be built in time to be of any use. If it cannot be built before the danger is past then the proposal is useless and uneconomic.

On 6 April Swan Hill aerodrome was confirmed an element within the total operation. As the Lake Boga facility proceeded a great number of official letters and signals bore the stamp SECRET. Secrecy was the order of the day, as RAAF Wireless Operator, LAC Bill Shanks, recalled:

> I was listed to fly from Cootamundra NSW on what was classified a "Secret Operational Flight". The aircraft was a twin engine de Havilland D84, our destination was Swan Hill, we had on board a high-ranking officer named Scotty Allan, (he had been a pilot with Kingsford Smith). We were not to divulge where we had been. Being a non-commissioned person I stayed at a boarding house [Coffee Palace, now the RSL Club] opposite the Railway Station, while the Commissioned Officers stayed at the Royal Hotel.

The concept of the Lake Boga operation was, at first, more extensive than that finally achieved. By 13 April, the following Department of Air, SECRET minute paper was circulated to a select few including those identified by code as S1, S2 and S5:

> DS. It is proposed to establish a flying boat repair depot at Lake Boga about ten miles from Swan Hill. Also located there will be a RAAF Seaplane Training Flight operating 4 + 2 Seagulls, and 3 + 1 Catalinas, and probably a NEI Seaplane Training Flight with 2 Catalinas, 5 Dornier DO.24s, 17 Ryan Seaplanes and 5 or more intermediate training planes. 2. Will you please indicate your requirements so that the necessary buildings and services may be included.
>
> 13/4/42 G Knox, Group Captain, DWB

By 28 April the Department of Air (DOA) advised that while the establishment of a Lake Boga flying boat repair depot would proceed, plans to incorporate a RAAF and NEI Seaplane Training Flight had been deferred. The document also set the establishment strength of No. 1 Flying Boat Repair Depot (No.1 FBRD) at 29 Officers, 82 Sergeants, and 719 Airmen. Included in the above figures were WAAAF personnel comprising 3 Officers, 1 Sergeant, and 108 ACW. The selection of a Sick Quarters site was now requested. An earlier DOI instruction read: "Install meteorological office and facilities".

A plethora of plans and paperwork now began circulating between bodies entrusted with establishment of the Lake Boga facility and the upgrading of the Swan Hill aerodrome. In addition to significant input by the Air Board, the RAAF, Qantas and the Civil Aviation Department, other agencies of the Australian Commonwealth and the Victorian state government were also involved, many newly created to administer elements required in the conduct of the war that was being fought on Australia's doorstep.

Around the nation and beyond, tens of thousands of matters military required urgent consideration, action and review. In the months and years ahead, decisions covering finance, building construction, plant and maintenance at No. 1 FBRD Lake Boga were regulated by communication between, and inspection by, the RAAF and members of government bodies including the following:

- The War Cabinet
- The Minister for Air
- The Air Board
- Department of Treasury, Defence Division
- Chiefs of Staff Committee
- Department of Interior, Air Services, Works Director-Air Services
- No.4 (Maintenance) Group RAAF
- Department of Interior, Allied Works

Department of Civil Aviation

Department of Air, Department of Aircraft Production

Directorate of Hiring (Army)

Division of Import Procurement

Building Standards Committee

State Rivers and Water Supply Commission (Rural Water Commission)

The Shire of Swan Hill

Initial funds for the establishment of the Lake Boga facility and upgrade of the Swan Hill aerodrome were included in a budgetary document dated 19 May 1942. In *Australian and American Works Projects, Appendix A* thirty-eight locations were listed through all mainland states, plus Port Moresby. An amount of £4,100,000 had been allocated for "RAAF Development Program: Provision of Aerodromes, Buildings and Services - Tentative estimated cost". The War Cabinet, together with the Business Board, approved funding of £150,000 to establish No.1 FBRD Lake Boga and to upgrade the Swan Hill Aerodrome.

Significantly, correspondence from the Chief of the Air Staff, Air Vice Marshal George Jones, regarding the Lake Boga project, stated:

> The work is necessary for the maintenance of operational aircraft, and therefore essential to the Defence of Australia.

While an earlier document indicated that the question of financial adjustment would be arranged between the Australian Commonwealth and United States Treasury authorities, an accompanying letter stressed:

> In accordance with policy, the estimates given provide only for the bare minimum facilities essential for operational requirements.

In the event a final amount exceeding £250,000 was spent on No. 1 FBRD.

Running parallel to these decisions were early moves to requisition land upon which the various elements of No. 1 FBRD would be constructed. In all, seven sites were required: Work Shop and Hangars, Stores Area, Living Quarters, Sick Quarters, Aid and Dental Post, Radio Transmitting Station and a VHF Transmitting Station.

As activities increased at Lake Boga, stock and station agent Percy Holland had hatched a scheme whereby he might add a little lustre to his reputation and perhaps acquire a number of sheep at a discounted rate. Accordingly, Percy had a telegram dispatched from the GPO Melbourne. It read:

> PT HOLLAND C/O POST OFFICE HAY NSW. BUY 2000 WEANER MERINO SHEEP BEST WARTIME PRICE STOP GENERAL DOUGLAS MACARTHUR.

By the time Holland pulled up in his old utility truck before the Hay Post Office to collect the telegram, he was a celebrity.

At Lake Boga, authority to secure right of entry and erection of buildings upon requisitioned land was "Pursuant to the provisions of the National Security (General) Regulations, Regulation Number R55, Priority A1, Priority Number 109".

Six acres were now requisitioned on the lake foreshore to the north of the township. Possession was immediate:

> Requisition of property, Lake Boga, Victoria. Property required for use as Flying Boat Repair Depot. Date required for possession IMMEDIATE, Priority A1.
>
> Signed RB Duncan Dated 1/4/42

West of the township, up on the rise along the Ultima Road, an extension of Lake Boga's Marraboor Street, an area of 50 acres was requisitioned for construction of living quarters. A stores area was finally sited on railway land adjacent to the Depot while the concrete signals, cypher and communications building within the Depot perimeter would be constructed as a semi-underground building.

Just east of the Lake Boga township eighteen acres was resumed for construction of a semi-underground transmitting station and towers. By May the planned requisition of land for a VHF radio station at Lake Baker was deferred. Installation of this facility did not proceed.

By the end of May, the DOI instructed that the semi-underground communications building should proceed, stating:

> The Building should be protected against bomb splinters and low flying front gun attack. It is therefore requested that provision be made for the erection of a reinforced concrete transmitting building.

Added to this resolution was confirmation to Qantas that an extensive area of foreshore adjacent to the Depot had now been acquired by the DCA for the establishment of a civil flying boat base and incorporating DCA's new mesh slipway. In order to ensure Qantas would have an adequate area in which to operate, supplementary plans were supplied providing for the diversion of the Murray Valley Highway to run west of its existing route, parallel to railway land.

Slipway No. 1 under construction. Interlocked steel piles driven into the lakebed surround the work site. In the foreground is a pontoon and jib used to position the 2-ton concrete mooring buoy anchors. (Neil Worner)

In point of fact, much of the Lake Boga construction work was now running ahead of Air Board approvals.

Visiting Lake Boga on 5 June, on his fourth and final visit, Vladimir Michels, met with RAAF officers P/Os Seddon and Bannister to finalise technical aspects of Michel's work. He then proceeded to Swan Hill where he thanked Shire Engineer Lowenstern for his assistance and departed the following day.

On 31 May Japanese midget submarines had raided Sydney harbour, and barely a week later Australians were startled when Sydney and Newcastle were bombarded by shellfire from a large Japanese submarine, causing superficial damage. Enemy submarines then attacked shipping along Australia's east coast. At Rathmines RAAF personnel had already dispersed flying boats around the perimeter of Lake Macquarie, backing them in under giant gum trees in an effort to avoid detection and damage in the event of an enemy air attack. It was at Rathmines that No. 1 Flying Boat Repair Depot was first formed 16 June with a strength of 24 personnel prior to its relocation to Lake Boga in July.

Neil Worner, SRWSC engineer, and his team arrived at Lake Boga 19 June entrusted with constructing the permanent concrete slipways, the Marine Section jetty, water supply and drainage to the Depot, camp and Sick Quarters, installation of chlorinators, erection of tank stands and radio masts, excavation for fuel tanks, installation of compressor lines, and the back filling and camouflage to the semi-underground communication building.

As State Rivers equipment was being set up, workmen were intrigued at the sight of a horse drawn caravan on the adjacent highway clopping slowly towards Swan Hill. Upon the sides of this caravan was painted "THE AIF NEWS". Workmen soon discovered that the caravan's owner was freelance journalist Alan Marshall. Marshall had one paralysed leg he had named "the Moron" and walked with the aid of crutches, calling these constant companions Isobel and Horace.

Among other work, Marshall had been writing humorous articles for the *AIF News*, a weekly paper published in

A hand operated pump supplies air to a Melbourne Harbour Trust diver who inspects a concrete buoy anchor installation. (Neil Worner)

Cairo for Australian forces serving in the Middle East. Marshall felt that a trip through the countryside gathering messages from parents, relatives and friends of servicemen abroad, would more vividly convey the news from Australia. In deference to the constraints of petrol rationing, he had a caravan specially constructed, mounting it upon the chassis of a T Model Ford, then purchased two horses he named Jim and Morgan. Together with his new bride Olive, Marshall had commenced his journey in early 1942, moving around the countryside enjoying the leisurely pace set by his horses.

Having passed Boga, by month's end the Marshalls were camped at Beverford and were befriended by local storekeepers Nell and Bill Athom who willingly assisted in gathering news of the district.

While at Beverford disaster struck. Marshall had gone to bring in the horses and had mounted Jim, bareback, by holding the horse's wither with one hand and using the top of a crutch with the other. Jim had suddenly pig rooted, dislodging its rider. As a result, Alan Marshall was admitted to the Swan Hill District Hospital with his already useless leg broken.

Flying Officer later Flight Lieutenant Gordon Myers the Depot's first engineering officer. Flight Lieutenant Myers served at No. 1 FBRD from 28 June 1942 until 2 August 1945, with brief visits to Morotai. (Gordon Myers)

June 28 saw the arrival of the first RAAF personnel to take up duties at No. 1 FBRD. They comprised Flight Lieutenant George Stewart Moffat, Temporary Commanding Officer, Flying Officer Gordon Myers, Engineering Duties, and their Transport Section driver, Flight Sergeant Colin Stewart. Flying Officer Myers would become one of the unit's longest serving officers.

By July 2 an advance party of 30 personnel arrived including Flight Lieutenant F Johnson, Barracks Duties. With the Lake Boga facility still in its construction phase, airmen took up residence in the Royal Hotel, Swan Hill. Flight Sergeant Colin Stewart recalled:

> Stewart Moffat, Gordon Myers and I picked up a new khaki Chevrolet staff car at the Melbourne depot on the last Sunday in June 1942 and drove to Swan Hill. On the days following, the officers arranged to take over the Royal Hotel and rented empty shops in the main street. The show room at Dowling's garage at the Boga end of the street became the Equipment Store, a garage next to the Commercial Hotel was rented, plus a shop next to Swinton's Furniture.

In addition, temporary headquarters were established at the originally numbered 91 Campbell St adjacent to the Commercial, now Westpac, Bank while a Barracks Store was set up at 83 Campbell Street, a shop within the Royal Hotel complex. Chisholm's Garage, the two-storey building opposite the Swan Hill Stores was now occupied by the Transport Section.

Mid-month, Flight Lieutenant EA Rowe arrived on posting from Bradfield Park, appointed the depot's first adjutant. Further officers to arrive in July were Pilot Officers W Wingrave (from Bankstown) and D MacGregor on posting from No. 4 Initial Training School, Victor Harbor, SA, for accounting and administrative duties respectively.

Lake Boga saw its first Catalina flying boat on Sunday 12 July when a quantity of stores and equipment was flown in from Rathmines. When compared to the Walrus that had alighted four months earlier, this second flying boat seemed enormous with its graceful fuselage and huge wingspan topped by two powerful motors. This sight was already familiar to many Depot personnel who had arrived on posting from RAAF Station Rathmines.

On July 22, No. 1 FBRD was declared a Master Stores Depot for Catalina Equipment. While RAAF chippies began construction of work benches and fittings, transport personnel journeyed to Melbourne returning in trucks laden with stores and equipment.

Hangar construction had just begun. The design adopted was that of an open fronted grandstand type hangar with a cantilevered canopy, "the eyebrow hangar". Eight large steel framed structures were erected, 120 feet wide by 58 feet deep. Of this depth 28 feet was a cantilevered roof section while the roof ridge rose 35 feet from the concrete floor. The Commonwealth Construction Corporation's sub-contractors had no such luxury as a crane with which to raise the hangar's steel stanchions and roof trusses but instead employed a 45-foot-tall gin pole, securely anchored at four points. When framework was complete hangars were sheathed in corrugated iron.

Hangar No. 1 framing up. Note the gin pole in lieu of a crane. (Neil Worner)

With beaching gear attached, flying boats would soon commence nosing into the first finished hangars. During the design stage, hangar width had been increased to ensure accommodation of wing spans of the RAAFs' largest flying boats, the C Class Empire with its 114 feet span and although still in Britain, Sunderland flying boats. However, these larger aircraft would not be beached at Boga.

Of this period, Corporal Richard Clapson, Stores Section, recalled:

> My pay book tells me I arrived at Swan Hilll on posting from Parkes NSW to No. 1 FBRD on 9 July 1942 and I left on posting to Morotai in the SWPA on 5 May 1945. The advance party of which I was a member was set up in residence at the Royal Hotel, and occupied bedrooms with two men to a room, and we ate in the hotel dining room. This was great, what a wonderful war, while it lasted.

In true military style, stores personnel marched through Campbell Street twice daily to and from their duties at Dowling's Motors Stores Depot. By this date, the tower atop Dowling's which had been erected in 1939 as an aircraft beacon, flashing SH in Morse, had been switched off as a security measure. Richard Clapson:

> Parades were held in the street alongside the Royal Hotel (McCallum St) - some men were taken to work by tender to Boga. The CCC were flat out building the base. The accommodation barracks and mess halls were all built as a cottage settlement, an extension of the Boga township to fool the Japs in case of an air raid. They were all weatherboard and cement sheet with iron roof construction and had no internal wall other than what was necessary to hold up the roof. Many buildings had fake chimneys.

On 22 July, six Wirraway aircraft from No. 7 Service Flying Training School, Deniliquin, called at Swan Hill in search of a Wirraway aircraft and its pilot, overdue on training. The aircraft and pilot were located later that day.

At this time, I recall witnessing a number of Wackett trainer aircraft at the Swan Hill aerodrome on a training exercise from Deniliquin. One of these bright yellow planes made too short an approach, striking the perimeter fence and overturned right beside me. Resolutely I crawled under the wing to help the stunned pilot out of his leaking aircraft,

Barwick's Royal Hotel, on the corner Campbell and McCallum Streets, Swan Hill. (Lindsay Denham)

The showroom of Dowlings Motors, on the corner of Campbell & Rutherford Streets was used by the RAAF as a temporary stores section from July 1942. (Norman Dowling)

only to discover how young he appeared. I anticipated that the RAAF officer who arrived would assist the shaken pilot to an adjacent tender and I recall being surprised when all he received was a severe reprimand.

No. 1 FBRD personnel had been in the area a matter of weeks when the Depot CO was approached by Swan Hill Town Clerk Jack Womersley who suggested that a Mayoral Debutante Ball be held with uniformed airmen as partners. This would provide the RAAF with an introduction into the community, and at the same time replace the shortage of local lads away at the war. Corporal Dick Clapson continues:

> They needed about twenty men. In true service fashion, on parade next morning twenty "volunteers" were nominated - you, you, and you, etc. My best mate and I were selected and he finally partnered a girl from Woorinen. My partner's name was Marj Ansell.

Towards the end of July, as personnel were organising basic elements of Depot establishment, the Japanese bombed Townsville on three occasions. The sole casualty during the three raids was one palm tree. The Japanese had slightly more success on 31 July when the Mossman area, north of Cairns, was bombed. Carmelia Zeillo, the two-and-a-half-year-old daughter of a cane grower was struck in the head by a bomb fragment. The child survived.

As author Marshall's broken leg slowly mended, he began working from his hospital bed. This unplanned incarceration provided him with an opportunity to describe the people with whom he found himself surrounded. By July Marshall had been moved to the hospital's flywire screened verandah with its flapping canvas blinds. Marshall enjoyed listening to the old Spaniard on his one side and an Aboriginal boxer on the other. In his book *These Are My People* Marshall wrote:

> Dick Ford, known as Whirlwind Ford, was a boxer. He was a member of a travelling troupe of fighters who toured the country shows for nine months of the year. In winter he cut wood and kept himself fit for the next season. He lived in a kerosene tin and hessian-bag hut on the banks of the river.

Whirlwind Ford loved everything, his family, his riverside camp, listening to the old blackfellows who visited him, and the Scots airman talking about his experiences (probably LAC Doug Tweedie). "I love listening to Scotchmen. Anyone who talks different to us is nice to listen to" Whirlwind told Marshall. Of his July convalescence Marshall wrote:

> The nurses were giving a party in their quarters. The airmen from the station not far away were invited, and I was included. The Matron gave me permission to go, and I got ready to attend my first party in pyjamas. When the night arrived two nurses wrapped me in blankets and lifted me into a chair. They carried me across the yard and into a spacious sitting room where blue uniformed airmen sat talking to girls I hardly recognised as the nurses who had been bustling through the wards an hour previously. They danced and sang, and I drank two pots of beer pushed surreptitiously into my hand by a generous airman. My condition, after nearly three months in hospital was hardly robust enough to meet the impact of alcohol on a convalescent stomach, and I began to wish for the solidity of my bed. It at least was stable.

With horses now out of the question, one was given to a farmer and Jim was sold at the Swan Hill sale yards. Marshall's Ford car with its modified controls was brought from Melbourne and would now tow the caravan. Resuming their journey, Alan and Olive travelled westward into the Victorian spring, covering as much territory as their monthly petrol ration allowed.

By August as workshops were progressively added to hangars, personnel had already converted the huge timber crates, arriving with aircraft spare parts, into temporary workshop accommodation.

Darwin had just sustained its 25th air raid when Lake Boga received its first Catalina for maintenance and repair, the flying boat alighting at 1700 hours on Wednesday 5 August. Having circled the lake and flown the descending legs, RAAF Catalina A24-17 came in gracefully on final approach, down to a copybook alighting. With props still spinning, canvas funnel shaped drogues were cast out of the blisters, arresting the flying boat's forward motion as a crew member sprang from the bow hatch releasing the mooring cable from the external anchor locker. With engines cut, seconds later, the Catalina was securely moored up to a large rubber buoy. The day was nearly done as Marine Section personnel motored out in the Depot's temporary launch bringing the crew ashore.

This weary war bird A24-17 had arrived in Australia in October 1941 when it was received by Seaplane Training Flight, Rathmines, just weeks before the Japanese attack on Pearl Harbor. Since January 1942, this flying boat had been flown relentlessly in operations against the Japanese by Nos. 11 and 20 Squadrons.

Catalina Squadron, First and Furthest, by Jack Riddell DFM AEA, refers to an important operation in which A24-17 was involved, sighting a major Japanese force headed for the Coral Sea:

> Cairns 5/5/42. Catalina aircraft A24-17 captain F/Lt F Chapman, second pilot P/O P Marsh to carry out a night reconnaissance of Rossel Spit, Laughlan Island, Woodlark Island, Jombard Passage, thence to Bougainville Passage. Aircraft sighted two enemy vessels and large aircraft carrier, two battleships or heavy cruisers travelling east, two submarines and one tender. Airborne 1655/L and return to Cairns. Flying time 19 hours 20 minutes.

Catalina captain Flight Lieutenant Mike Seymour and his crew later staged to Noumea, from where they flew a 17 hour round trip to bomb the Japanese seaplane base at Tulagi. On the homeward leg at 0800 hours, as Seymour commenced his descent to Havannah Harbour, Efate, the Catalina was suddenly strafed from above by a US Marine Corps F4F-3 Wildcat fighter. The Wildcat's pilot was taking no chances. Seeing a large camouflaged flying boat heading for the US base and displaying bright red roundels, the outer white and blue circles apparently appearing merged with the camouflage, the US pilot attacked. The RAAF Cat held its fire until recognition was established, but A24-17 had been hit during each of three diving attacks. Flight Lieutenant Seymour later described the damage:

> Both fuel tanks were holed, the port aileron was shot away, the hull was holed in seven or eight places. One round passed down the back of co-pilot Sergeant Brammel's seat and into the bilge, from where he later

A24-17 undergoing overhaul at Lake Boga. The red centre to the roundel, deleted soon after, is still evident at the time this photograph was taken, in August or September 1942. (Lindsay Denham via David Vincent)

retrieved it.

After this incident, the red centres were removed from the roundels of all RAAF aircraft serving in the Pacific and replaced by a blue circle surrounding a white centre. American aircraft, having also encountered friendly fire, had earlier ordered removal of the red "meat ball" from within the centre of their five-pointed white star on its blue field.

After repairs aboard a US Navy seaplane tender the flying boat returned to operations before it was flown south to Lake Boga for its 240 hourly inspection and thorough restoration.

On arrival at Boga, Catalina A24-17 carried a crew of four plus twelve personnel on posting from Rathmines. Immediately passengers were ashore the aircraft was slipped from the buoy and taxied to the lake shore where beaching gear was fitted. The gear consisted of pairs of wheels that were attached to either side of the fuselage between the wing struts and a rear steerable wheel on an inverted tripod. A Depot tractor towed A24-17 up the DCA mesh matting, entering the Depot proper through a side gate in the barbed wire perimeter fence.

Among the new arrivals was Fitter IIA, LAC Jack Settle who recalled:

Men of the No. 1 Flying Boat Repair Depot RAAF, practise their "small boating" skills at Lake Boga in August 1942. (AWM)

An RAAF Catalina alights at Lake Boga in 1942. (AWM)

> I came down on the first Catalina ever to be repaired at Lake Boga, it had recently been shot up. Beaching gear was fitted out in the water and a tractor towed the plane out over heavy mesh matting. There wasn't much there at Boga. An old farmhouse still stood in the Depot area. Hangars were started and a slipway. No one had ever heard of Lake Boga. Our aircraft was packed with personnel from several musterings, Fitters IIA and IIE, signals, fabric workers, an engineering back up, and stores personnel.

A description of the Consolidated Aircraft Corporation's PBY 5 Catalina and its distribution of RAAF aircrew appears in Jack Riddell's *Catalina Squadrons First and Furthest*. The following is based on that account.

With a wingspan of 104 feet, the Catalina's fuselage was 64 feet long. The height of the tail, over 18 feet. The flying boat was powered by two Pratt and Whitney, twin row fourteen-cylinder R-1830 radial air-cooled engines, each of 1200 HP. Recommended weight on take-off was 35,000 pounds including 1,460 gallons (1,750 gallons US) of petrol, 4,000 pounds of bombs, six machine guns, and ammunition. Operational RAAF Catalina crews consisted of nine personnel: the captain, second pilot, navigator, first engineer, second engineer, 1st wireless-telegraphist (W/T), 2nd W/T, rigger and armourer. Prior to March 1942, RAAF Catalinas had flown without the assistance of navigators. The first engineer was directly responsible to the captain for the serviceability of the aircraft and all other matters related to the organising of the crew.

Catalinas were divided into seven compartments, with six watertight bulkheads. The bow compartment was used by the bow gunner (2nd W/T), and when bombing, by the bomb-aimer. Next was the pilot and co-pilot's compartment with its large inverted U-shaped control yoke under which access to the bow could be gained.

The third compartment accommodated the navigator, wireless operator and radar operator, while the fourth space under the mainplane housed the galley equipped with a two-burner electric hot plate and an auxiliary power unit which supplied power for waterborne radio transmissions, energising starter motors, and bilge pump operation. Immediately above the fourth compartment was the "tower" configuration that connected the fuselage to the mainplane. This was the engineer's position, a forward-facing space, a window to port and starboard plus a large control panel. Instructions from the Cat's captain to the flight engineer could be received by both headphones and a series of coloured lights, switched from the cockpit control saddle. Immediately above the engineer's head, in the central wing section, were the fuel tanks. The bunk compartment held four double decker bunks. The blister compartment carried a machine gun to port and starboard. The final space was the rear tunnel compartment housing one machine gun, the flare chute, and a portable head (toilet).

Around the interior of the thin-skinned aircraft was a broad red line, the Plimpsol mark or waterline of the aircraft when unladen, providing crew with a guide to the aircraft's floatation if shelled or damaged.

Swan Hill's Flight Lieutenant Dr FE "Dick" Browne, No. 1 FBRD's first medical officer. (FE Browne)

By the time Catalina A24-17 arrived at Boga, 30 such Consolidated PBYs had been taken on charge by the RAAF. Of these 27 had been ferried out new from the USA, while two were ex-US Navy Cats and one, ex-NEI. Of the 30 flying boats, twelve had been lost, together with the lives of 49 RAAF and two attached US aircrew. Two of the twelve Catalinas lost had been confirmed shot down by enemy action, with three more presumed to have met the same fate. Five Catalinas had, while beached or waterborne, been destroyed or damaged during Japanese attacks on Port Moresby and Tulagi. One flying boat had crashed on take-off at Port Moresby, while one had been lost in a heavy swell at Kavieng.

On 21 August, Air Commodore EC Wackett and party arrived at Lake Boga on inspection. Although No. 1 FBRD was administered through No. 4 (Maintenance) Group, HQ, Melbourne, Air Commodore Wackett was the Air Member for Engineering and Maintenance, having ultimate engineering responsibility. The name Wackett was by this time synonymous with both the RAAF and the Australian aircraft industry. EC Wackett's brother, Lawrence J, was the driving force behind the 1936 establishment of the vast Commonwealth Aircraft Corporation. The Wackett trainer was an aircraft from the CAC stable.

August saw Swan Hill's Dr FE "Dick" Browne appointed the Depot's first medical officer. During the First World War, Dr Browne had served in both the merchant navy and RAN and now transferred from the army to the RAAF. He was known for his swarthy good looks and the staff at the Swan Hill District Hospital called him Charles Boyer Browne.

Swan Hill, early morning. In her mannish grey suit, vegetable grower Dorrie Le Bache was watering her horse at the Bills' horse trough in Campbell Street's intersection. Sharing a great love of horses, Annis Bills and her brother George had funded the provision of horse troughs in many communities throughout Australia. Over by the Royal the RAAF had assembled at morning parade, out on an unsealed shoulder of McCallum Street roadway. Roll call done, an Orderly Officer read daily routine orders after which the airmen were dismissed, dispersing to their varied duties.

Charlotte "Nell" Frazer who worked across the street at the Post Office telephone exchange recalled this period:

> The base at Lake Boga had a big impact on our working lives, we had earlier taken over the night shift to relieve the male night attendants for war duties. After lessons learnt in the first attack on Darwin, an air raid shelter was excavated in the back yard of the Post Office. Our windows were blacked out at night and security precautions were stepped up even to signing a book that we had locked windows and doors. Each morning a Fixed Priority Call was booked from 9.00 am to handle calls between No. 1 FBRD and Southern Command, Melbourne. The calls lasted up to 45 minutes.
>
> In those days all civil trunk line calls had to be booked and a time delay quoted. It became very difficult to quote customers the delay. We worked three and four hours without a break. The evening traffic was very heavy. After 9.00 pm the charge to Melbourne was 1s 3d for 3 minutes. All calls from servicemen were charged at half price, we took their name and service number, with their calls limited to six minutes.

Depot Area No. 1 Flying Boat Repair Depot

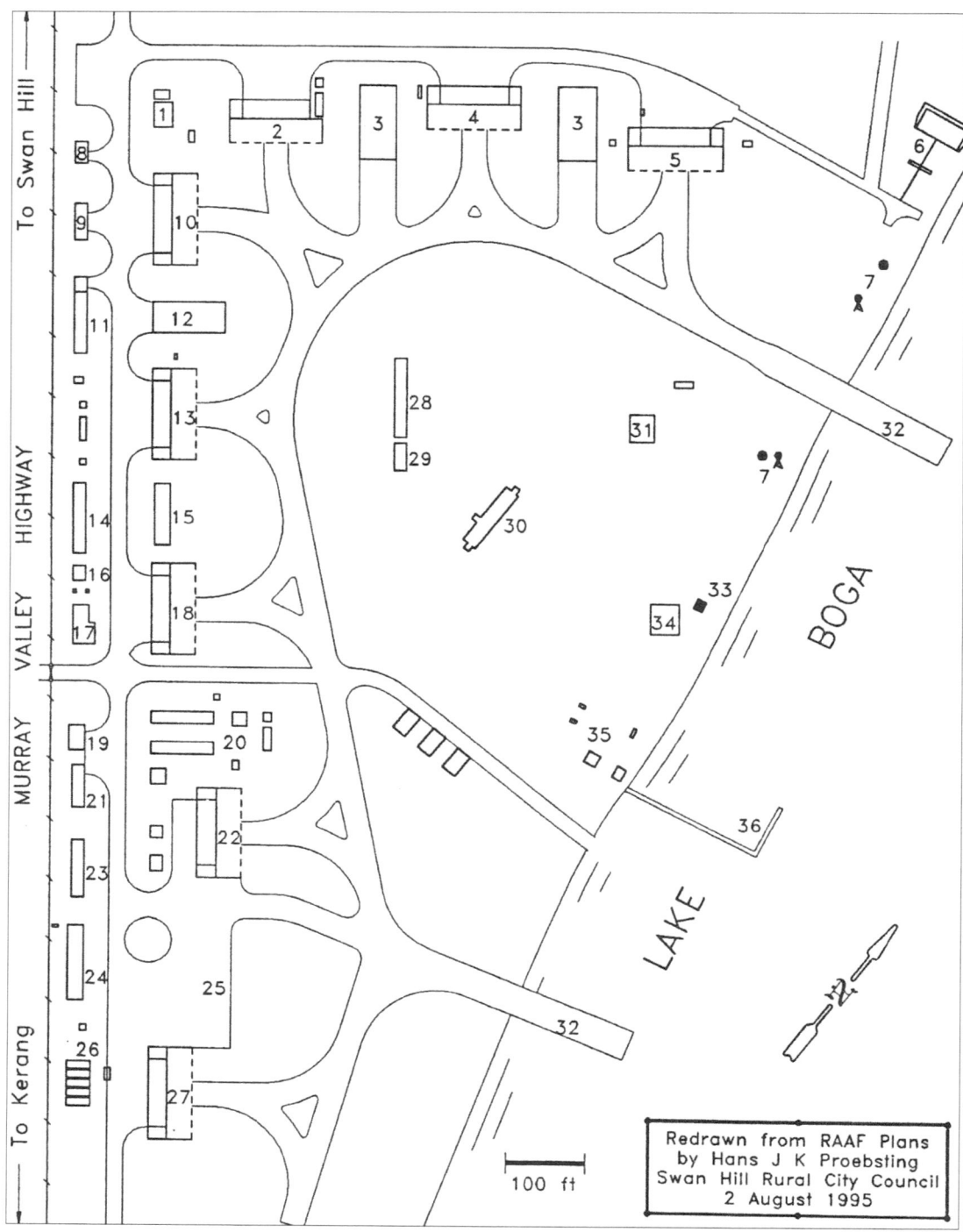

Legend
1. Emergency Power House
2. Woodwork Shop
3. Airframe Repair Shop (x2)
4. VP Propeller Test House
5. Paint, Dope, & Fabric Shop
6. 25-yard Machine-gun Range
7. Windmill & Tank (x2)
8. Inflammables Store
9. Oil Store
10. Engine Installation & Hydraulics Shop
11. Clothing Store
12. Store
13. Metalwork Shop
14. Electroplating Shop
15. Heat Treatment & Salt Bath
16. Blacksmith Shop
17. Guard House
18. Machine Shop
19. Garages
20. Headquarters & Administration Area
21. Battery Charging Room
22. Instruments, Wireless & Electrical
23. Workshop
24. Armament Repair Shop
25. Marine Craft Apron
26. Vehicle Fuel Point
27. MIT & Armament Repair
28. ARS & GES Headquarters
29. Photo Hut & Drawing Office
30. Signals & Cypher Room
31. Guard Quarters SP
32. 40 foot Slipway (x2)
33. Duty Pilots' Tower
34. Marine Store
35. Crew Rooms
36. Fuelling Jetty

By September, all manner of machinery was arriving on site, including an electric swageing machine that melded metal to metal when joining turn buckles to stainless steel control cables. Electro-plating equipment and a large guillotine now arrived at the rail head, while the Major Furnace Company had delivered a furnace and salt bath, which would render duralium sheeting more malleable after immersion. Transport Section vehicles were increasing in number with the arrival of an International flat top, a GMC fuel tanker, a Dodge utility and a light tip truck. An Indian motorcycle completed the inventory at this stage.

The Royal Hotel had now become over-crowded and additional sleeping accommodation was sought. The solution adopted was the erection of tents on the spacious lawns of the Swan Hill Club, the gracious former residence of W Garden. RAAF personnel still messed at the Royal Hotel.

At Boga the throb of multi-motored aircraft drew marked attention on 2 September, as the unfamiliar silhouettes of two Dornier Do24K flying boats grew more distinct. Four hours 20 minutes out of Rathmines, the first such craft alighted as Squadron Leader Gordon Stilling brought in Dornier A49-3 which was immediately followed by Flying Officer Norm Fader aboard A49-5. Each RAAF Captain commanded a crew of four.

The only flying boat in WWII to be flown by both the Axis and Allies, the Dorniers were of German manufacture, to a Dutch order for use as a military flying boat for surveillance of the thirteen thousand islands and territories comprising the NEI, an extensive colonial empire. Thirty-six flying boats out of an order for seventy-two such Do24Ks were delivered to Java prior to the outbreak of war in Europe. During the defence and withdrawal from Java, the Dutch lost most of their Dorniers, many by Japanese air attack. While twelve Dorniers reached Australia, six were destroyed during the disastrous 3 March Broome attack. Now just six Do 24Ks remained.

By now the Dornier Do24K flying boats were rare birds indeed. These aircraft were equipped with three Wright Cyclone 875 HP engines mounted high on the mainplane and had a slim fuselage which swept back to support a twin finned, monoplane tail. Sponsons or "sea wings" at the fuselage water line provided lateral stability, negating the necessity for floats. The sponsons also carried the aircraft's fuel supply. By now they had been stripped of their armament and were used by the RAAF as transport and communication aircraft. When earlier operated by the Dutch, the Dorniers had carried bomb racks with machine guns in the bow, waist and tail.

Having arrived in Australia with their Dutch registration prefixed "X", five were purchased by the RAAF in June 1942 when their registrations were changed to the RAAF nomenclature of A49-1 through to A49-5. A sixth Dornier, X-24, initially used by the Dutch, was transferred to the RAAF in October 1943, then becoming A49-6.

RAAF Dorniers were first allocated to No. 41 Squadron in August 1942 but with difficulties encountered during restoration they would not join the squadron until mid-1943. Depot personnel now began grappling with the task of working on these unfamiliar aircraft. With beaching gear not yet available, large bays were dug in the lake foreshore to accommodate Dornier bows and sponsons. Fitters IIE and IIA, electricians and instrument section personnel now worked on the Do 24s under extraordinarily primitive conditions prior to the aircraft being beached and hangared.

Depot strength was building. During September, Coleraine grazier LAC Ellis Bickley was posted to Lake Boga from No. 1 Wireless and Gunnery School, Ballarat. Bickley recalled:

> I missed out on aircrew because of a suspected hearing defect and was posted to No. 1 FBRD on 3/9/42. I arrived by train and reported to the Royal Hotel. A small building in the main street had been taken over as an armament workshop, and the Sergeant in charge of our section, Joe Finkelstein, took me on as an armourer's assistant. After passing a trade test, I became an armourer.

> Supplies of .5 calibre ammunition from the USA were no problem, however there was little for the .30 guns. Our task was to strip the .30 Brownings and convert them to .303 calibre ammunition, readily available in Australia. We replaced the .30 barrels with .303 barrels, but because all American ammunition was rimless and our .303 bullets had rims, the breech blocks had to be changed too.

LAC Bickley also noted how primitive initial Depot conditions were and the particular problems encountered with the overhaul of the Dornier flying boats because no beaching gear was yet available.

The Dorniers were towed into a water filled trench, dug into the bank of the lake, and the engine and airframe fitters had to work out in the open or in rough makeshift shelters on site. The Dorniers presented special problems in that they had large midship waterline floats as part of their hull buoyancy, so special wide trenches had to be cut into the lake's bank for them.

I remember Victoria had 6 o'clock closing of hotels, the Federal Hotel or "Bloodhouse" across the river in New South Wales was popular with those seeking alcohol. Beer was sixpence a glass or if you were short of funds, a "Ginger Meggs" [after the Sun News Pictorial comic strip character], a mixture of sherry and dry ginger ale, cost fourpence. I believe a hangover from the latter was something to remember.

Early days. An RAAF parade in partially sealed McCallum Street, Swan Hill. The English, Scottish & Australasian Bank is in the background, now ANZ Bank. (Bill Muller)

A Dutch Do24K tri-engine flying boat, six of which arrived in Australia and served with the RAAF. (AWM)

On September 12, the Depot's first serviced and repaired Catalina was about to depart in a night time take off. His landing lights ablaze at 2345 hours, Flight Lieutenant Bob Hirst with a crew of ten departed Boga in Catalina A24-17, the first of many hundreds of restorations that would be carried out in the years ahead.

Friday 25 September saw the arrival of Dornier A49-4 and Catalina A24-26. Operations in which the Catalina had recently been involved included a rescue mission to northern New Guinea.

As the first group of buildings neared completion at the living quarters, Lake Boga residents now began to appreciate the magnitude of this sudden wartime intrusion upon their tranquil township. RAAF structures would soon outnumber local buildings. Lake Boga resident Eddie Scown recalled:

> My parents had a fruit block at Tresco West. My father, Charlie, also drove a tip truck for Jim Baker carting sand to build the RAAF slipway. Each morning I drove a horse and cart up to the camp and delivered building materials from the big fenced in compound, to the various contractors the CCC had working for them.

The Lake Boga Mechanics' Institute committee was taking no chances when, at its monthly meeting a motion was passed to take out special War Damage insurance of £2,000 on their Marraboor Street building.

September had seen temporary CO Flight Lieutenant GS Moffat receive promotion to squadron leader and at this time ex-No. 10 Squadron's Flying Officer Jack O'Donnell arrived. In October, Depot development advanced when electric power was permanently installed, the Signals Section received major items of radio equipment and the Marine Section at last had its first permanent service motor launch.

As work continued with construction of WAAAF accommodation, 150 airmen vacated the Royal Hotel Swan Hill and moved to the Depot living quarters. Each airman was issued with a large hessian envelope which when filled with straw became his palliasse. In addition to those now domiciled at the camp, married airmen were permitted, with their wives and families, to live out at Swan Hill. Several personnel, among them Flying Officer Gordon Myers, rented a Federal diesel truck from transport operator Charlie Wendell. This vehicle proved a reluctant starter requiring a bonnet up application of a cloth and ether to fire up the motor; however, the vehicle proved invaluable.

Together with a small number of airmen returning from Britain having served with the Sunderland equipped No. 10 Squadron, RAAF, No. 1 FBRD personnel were drawn from units and training centres in most Australian states and by November, postings brought the Depot to near establishment strength. Upon notification of a posting, personnel turned immediately to an atlas or road map. Where the devil was this place called Lake Boga?

Most would journey to the new RAAF base by rail via Melbourne, passing through the old gold mining city of Bendigo, then on to Kerang. Maps indicated the final leg of the train journey would travel through Fairley, Lake Charm, Mystic Park, and Tresca, then to Lake Boga. What a posting! To some, the very sound of these place names conjured up scenes of the luxuriant Cyprus Gardens, Florida, USA, as portrayed in James Fitzpatrick's movie travelogues. Those from greener pastures were in for a surprise!

As the newly posted alighted at the Lake Boga Railway Station, there was just time to survey their surroundings before tenders transported them up to the camp. From the platform, eyes were drawn across the highway to the lake, an inviting expanse of water with a treed public foreshore plus a collection of modest lakeside cottages. To either end of the rail head, the shining silver track divided an expanse of stunted Yanga salt bush and parched red sand. Away from the lake and across the railway line, stood a small township, the occupants of which had already begun to extend the warmest of welcomes to RAAF personnel.

From their canvas covered tenders on their way to the camp, new arrivals had but a tunnelled vision of Boga's Marraboor street. At the edge of town, the RAAF vehicles roared up the road, then slowed on reaching the barren limestone rise, upon which rested elements of the new RAAF camp.

Crows carked from the branches of a gnarled box tree by McGrath's slaughter yard some way distant, as newly arrived personnel surveyed the surroundings that were to become their home. At that moment many may well have paraphrased the Swan Hill High School's theme song, had they known the words.

> On an ant hill near a sand hill in the Mallee bare and wild, stands an air base, yours and my place, and our time is there beguiled.

After five weeks absence, Catalina A24-17 was back for an engine change as No. 11 Squadron's Flying Officer "Bimbo" White with Flight Sergeant Bob Honan second "dick", alighted on 22 October. Safely down, the crew transferred to the Royal Hotel where Flight Sergeant Bob Honan found the resident mosquitos "as large as Catalinas and far more ferocious". In his book *That's That* Bob Honan reflects on his first flight to Lake Boga:

> Boga was difficult to find because as you approached the River Murray there were many tributaries and back waters which made map reading difficult.

The term, second "dick" or "dickie", as applied to RAAF second pilots was in frequent use having been acquired from the RAF. The use of the word dick originated from the days of the horse and carriage when drivers known as dicks rode in the open, up on the dicky seat, the assistant driver being a second dick. Similarly, the fold out seat to the rear of early motor vehicles was known as a dicky seat.

On his second visit south, Squadron Leader Gordon Stilling alighted from Rathmines on 19 October in A24-19 with his passengers, the US Navy's Lieutenant Commander Schoech and Group Captain Poole. Once ashore the officers proceeded to inspect the RAAF Unit; however, nine months would pass before US Navy aircraft would arrive at Lake Boga.

Lakeside dwellers and bird life had their sleep disturbed at 2200 hours that evening when Squadron Leader Stilling powered his Catalina across the shadowy waters on a return flight to Rathmines. As October closed, the State Rivers' Harman dragline was hard at work extracting the metal piles from around the completed slipway and a SRWSC team had already started driving timber piles at the Depot jetty site.

The relentless efforts of RAAF Catalina Squadrons over preceding months had not gone unnoticed. US Major General George C Kenney, Commander Allied Air Forces South West Pacific Area, advised:

> I have carefully noted the excellent work accomplished by the Catalinas during September 1942. Despite bad weather which prevailed, particularly during the latter part of the month, your squadrons compiled an enviable record by completing missions on 29 days.
>
> The attacks resulted in heavy damage to enemy aerodromes and installations at Buka, Kahili and Buin as well as shipping in the Buin-Faisi area.
>
> The courage and determination with which the numerous operations were carried out during the period were a contributing factor to the success of our combined Allied forces.

With weather warming, physical training was conducted on the camp parade ground as personnel turned out in summer khaki drabs and sand shoes. As the district's summer dust problem emerged, several fly wired kitchens and food preparation areas at the camp were glazed.

Flight Lieutenant "Bimbo" White, Flight Sergeant Bob Honan and their crew flew north again, on Friday 30 October, aboard Catalina A24-17. By the following Sunday night they joined Nos. 11 and 20 Squadrons in a six Catalina bombing strike out of Cairns against Buin and Faisi. Bob Honan recalled:

> The weather was bad until we approached the target area at 0300 hours. The main target was the Japanese shipping in the harbours, and at that time, there were as many as 100 ships in the vicinity. It was a tense time as we approached the target. The crew donned parachutes and placed the .38 pistols in their holsters with emergency rations etc in case of aircraft abandonment. The bow gun turret, the rear tunnel gun hatch and the blisters were open with the guns ready and manned. There was intercom between all positions and a warning horn for various signals. We dropped our bombs as a stick, that is one after the other in a line but did not observe any hits, and the ack ack was relatively quiet. However, seven enemy fighters were sighted by the crews of some of the other aircraft.

Following a safe return to Cairns on Monday morning, the White-Honan combination was rostered to fly again on Wednesday on a 15-hour night strike against the same Solomon Island targets. After delivering their 6 x 500-pound bombs they returned through a bad storm. Upon alighting back at Cairns on Thursday morning, they found a suspected fuel tank leak. Accordingly, A24-17 was again flown to Bowen and Lake Boga. Flight Sergeant Honan:

> On Saturday 7 November we departed Bowen early for Lake Boga, with a slight diversion on track to drop S/Lr Chapman's mail at Charleville which is 500 miles from the coast. It was my first experience of flying over the dry, arid inland of Australia with mile after mile of red landscape. At Charleville we put the wing tip floats down and lined up as if to land down the main street. Some of the locals must have got a surprise.

Having alighted at Boga, Sergeant Honan, his skipper and crew were billeted on this occasion at Swan Hill's White Swan Hotel where they were courteously received by the young proprietor June Wilkins.

The first W/T on the White-Honan A24-17 at this time was Kerang resident Flight Sergeant Lindsay Rundle DFM who took this opportunity of spending a brief leave in his hometown. Lindsay had 1,175 flying hours under his belt having flown on Catalinas since April 1941 during the constant surveillance flights from Port Moresby, to Rabaul, Tulagi, Vila, and Noumea, and then covering the route in reverse.

By November, a survey of emergency alighting areas between Rathmines and Lake Boga was carried out. Walrus W2755 now arrived with its survey crew. In addition to the identification of suitable bodies of water, farmhouses with telephones adjacent to emergency alighting areas were recorded with lakes and homesteads in the Balranald district included.

On 2 November, WAAAF squadron officer, Audrey B Herring, arrived at Lake Boga on a two-day inspection visit of the RAAF facility. This visit was an indication that the posting of the first contingent of Women's Auxiliary Australian Air Force to No. 1 FBRD was imminent. By month's end, Assistant Section Officer Sheila Patrick arrived on posting, to take up duties as the Senior WAAAF Officer, Administration. Prior to enlistment, the attractively red haired ASO Patrick had been a journalist and was a member of a family long associated with the Australian coastal shipping industry. Within thirty days of her arrival, ASO Patrick would have 37 WAAAFs under her direction and at all times would do her utmost to successfully integrate this new element into a cohesive force, working side by side with Depot airmen.

November 16 saw the temporary commanding Officer Squadron Leader George Moffat relinquish command of No. 1 FBRD to Squadron Leader Geoffrey Douglas Marshall who arrived on a posting from No. 5 Aircraft Depot in Wagga. A graduate in mechanical and electrical engineering from the University of Western Australia, Squadron Leader Marshall had joined the RAAF in 1938.

Until November, test flights within the Lake Boga region had been relatively disciplined. Then it started! The terrifying yet exhilarating sight and sound of huge flying boats "shooting up" the local towns, coming from nowhere then suddenly thundering overhead, their great shadows flicking past in a mini second. During test flights and when departing the area, Catalinas now regularly indulged in low flying over Lake Boga and Swan Hill at what appeared as tree top level.

While many residents were enthralled by this daring spectacle, not so "Bluey", an airman from the signals section, who wrote to the *Guardian*:

Squadron Leader (later Wing Commander) Geoffrey D Marshall who served as commanding officer, No. 1 FBRD from 16/11/42 until 29/9/43. (GD Marshall)

> THE HEIGHT OF IDIOCY Plane Stunting Over Swan Hill
>
> On Wednesday afternoon a seaplane flew in a reckless manner round the town several times culminating in wild leaping stunts from tree to tree and a game of leaping from building to building in the main street. If the pilot has no little regard for his own life, he might at least consider the danger threatening the lives of his fellow passengers, also the disaster a crash at such a low level could have caused to individuals and buildings, not to mention the serious loss of a seaplane which cost thousands of pounds to build.

Public criticism of the RAAF by its personnel was forbidden and Bluey was soon charged then incarcerated in the Swan Hill lock up; however, all was not lost. A keen Morse code exponent, the airman was supplied with a Morse key by Flight Sergeant Laurie Benham and profitably whiled away his time.

The low flying was reported in more jocular fashion by the editor of the Swan Hill Rotary Club's weekly news bulletin, Shire Secretary, Frank Womersley, when he noted:

> It has been ascertained that the daring young pilot who caused so much consternation a few days ago with his house hopping, was really not so daring after all. In actual fact he was just a novice flyer and was frightened to go up any higher.

The low flyers were subsequently traced in the pages of flight engineer Jack Riddell's logbook. Jack was the engineer aboard Catalina A24-27, when flown by Flying Officer Clem Haydon during a one hour 25-minute test, on 4 November. The low flying was repeated on the following day.

In *Catalina Squadrons - First and Furthest*, Flying Officer Jack Riddell, DFM, AEA wrote:

> The aircraft carried as guest one of Swan Hill's leading medical men. During Clem's banks, climbs, turns, and dives the doctor fell into the bilge which resulted in a broken ankle. Jack Riddell had overlooked having the doctor sign a form absolving the RAAF from any claims should an accident happen. After making him [the doctor] comfortable on one of the upper bunks he happily signed the release.

A Catalina "test hops" down Campbell Street, Swan Hill. Low flying by Allied pilots was common practice after repair and restoration of flying boats. A composite photo. (Brett Freeman)

Just as the Lake Boga community of just over two hundred residents required its own police station and police officer, so too did the new RAAF establishment. With a projected strength almost five times that of Boga's population, the RAAF already had its own Service Police and now awaited construction of a lock up attached to the guard house. The duties of RAAF Service Police were to maintain good order and ensure the dignity of their branch of the services. These supervisors of RAAF propriety became known as SPs, the same acronym given to Australia's illegal starting price bookmakers. At this time, a newly posted young airman, hearing constant reference to SPs, discreetly enquired which airmen he might approach in order to put a few bob on the Melbourne Cup.

A month to Christmas and Catalina A24-1, captained by Squadron Leader Tom Stokes with sixteen airmen on board alighted on the aircraft's second visit to the lake. Flying second dick on that occasion was Flying Officer Bert Delahunty whose log recorded:

> 26/11/42 - PBY A24-1 S/Ldr Stokes, self and crew. Rathmines to Lake Boga and return 8 hours. Navigation exercise for new crew and deliver spare crew to collect a repaired aircraft.

At Lake Boga in late November, LAC Ron Monk moved out from the massive timber crate that temporarily housed the instrument repair section. Immediately he detected an unfamiliar stillness and instinctively looked toward the Depot's western perimeter. The sight that greeted him was awesome. RAAF personnel were about to experience their first Mallee dust storm.

A silent wall of dust, some 2,000 feet high, was rolling relentlessly towards the lake on a vast front, its graduated layers of tan, red and brown soils, deepening progressively from the storm's crest, down to its thick and foreboding dun coloured base. Suddenly the storm swept over the adjacent sand ridge and was down upon the Depot. A sensation of a thousand needles stung the hands and faces of unprepared personnel as they ran to batten down unsecured material and tighten moorings on rocking Catalinas standing unhangared on the Depot roadways. Waterborne aircraft reared and tossed at their moorings but remained securely restrained.

In thirty minutes the worst was over, giving way to a secondary cycle of cooler winds that bore a dust haze until dusk had fallen. Local residents measured this storm in the middle range: it had only been a semi-blackout. Personnel now faced the task of cleaning the Depot, its aircraft, and a camp all resting deep in Mallee dust. This spectacular natural phenomenon was a forerunner of storms that were to beset the area in the drought-driven years ahead, severely testing the resolve of both district residents and Depot personnel alike.

Pere Robbie was Lake Boga's nightman and as he moved through the streets aboard his horse drawn, twenty-four-cylinder night cart his one-armed assistant Hughey Stewart roved through Boga back yards performing his valued service. Robbie could not believe his luck when he got wind of a belated direction from the Department of the Interior dated 14/11/42 advising:

> In view of there being no waterborne sewerage available, a pan system would be introduced.

Things were already booming. Imagine, a guaranteed five hundred percent increase in business. Pere and Hughey would now have their work cut out, but then Hughey was most adept at his job as assistant nightman for he found the hooked prosthesis that replaced his lost arm just the thing for grasping a can handle and swinging it up onto the cart.

The Depot's strength at November's close numbered 604 personnel comprising 19 officers, 584 airmen and one member of the WAAAF.

December 1 saw the arrival at Swan Hill of Avro Anson AW664, when Squadron Leader Jones visited on investigation of ambulance fittings to Dornier aircraft. Three days later Squadron Leader Reg Burrage DFC MID (ex-No. 10 Squadron) alighted in A24-1 and was fascinated by the numbers of tortoises observed coming up for air in Boga's clear waters. In discussion with Depot personnel, Burrage was assured the reptiles were "tortels" a breed unique to the area. Enough said!

The guard house and the SP lock-up adjacent to the Depot gate were now complete as was a basic bungalow at the camp site which would provide alternate aircrew accommodation. The State Rivers team had begun the erection of three radio masts surrounding the Communication Building, each 70 feet high. After similar towers were erected at the Radio Transmission Station east of Boga, RAAF personnel strung the towers with appropriate cables.

On Saturday afternoons things could become pretty boisterous at the Boga pub as Neil Worner recalls:

> While engaged at No. 1 FBRD I stayed at the Boga pub and was walking through the hotel hallway one Saturday evening when I heard screams from a distressed female coming from the bathroom. Calling through the locked door I quickly established the occupant was the hotel proprietor's wife, Mrs Ryan. Observing propriety, I summoned Doc Ryan from the bar. He immediately broke through the door to find his wife near disappeared. The bathroom floor had been white-anted and with the weight of Saturday's bath water, the floor had completely given way. Apart from a bruised composure, Mrs Ryan emerged from the experience quite unscathed.

Brian "Tubby" Higgins of Wangaratta flew A24-2 in for an engine change and 240-hour overhaul on 7 December. Among operations flown by Higgins aboard this aircraft had been an unusual rescue mission on 19 September, to the Gulf of Carpentaria in search of the missing aircrew of a USAAF B-17 Flying Fortress. David Vincent describes this operation in his book, *Catalina Chronicle*:

> Parachutes were found tangled in trees, these men later being located by blacktrackers. A further four members were located by Higgins in the late afternoon of 20 September on a salt pan. Higgins landed on what appeared to be a coastal river but by next morning when the aircraft was to return to Cairns, A24-2 was high and dry on a mud bank, the tide having receded. By 1240 the return of sufficient water enabled the Catalina to successfully take off with the four extra personnel.

By now, WAAAF personnel were settling into their new surroundings and making the best of their austere accommodation. Domiciled in their cottages each WAAAF had been allotted a folding iron framed bed, blankets, a palliasse and a locker. With a public roadway running through the camp separating the sexes, one of the first signs erected outside WAAAF quarters read: OUT OF BOUNDS TO ALL RANKS EXCEPT WAAAF. While this was standard procedure for WAAAF sleeping quarters, during the day airwomen worked alongside airmen on similar tasks.

Every RAAF Unit had its budding poets with Lake Boga no exception.

MAGNIFICENT OBSESSION

Joining up to do a job
Releasing men to fight
Learning things they never knew
Almost over night

With Tojo almost at our door
WAAAF soon called to recruits,
Out of civies, into jeans
And airmen's modified suits

The Depot's first Christmas Day provided personnel with an opportunity to ease down a little as a traditional dinner was served within the sparsely decorated messes. No such luck for Flight Lieutenant Terry Duigan who was airborne out of Cairns aboard A24-21. Regarded as a free spirit, Terry was determined not to spend Christmas without his wife Gwynne. He had her on board his Catalina. Gwynne Duigan:

> When I left the Cairns wharf for the flying boat I was wearing a neat navy dress, I think onlookers thought I was some sort of special agent. When we arrived at Swan Hill the lady who ran the Royal Hotel [Nana Barwick] was most kind to us. That flight down from Cairns was not my first in a Catalina. Terry once took me on a test flight out of Bowen, I was in the co-pilot's seat. He said, "Have a fly!" I was astounded how heavy the controls were, the nose was either up or down. I found it difficult to keep the horizon where Terry told me.

The year almost concluded, low flying continued with local inhabitants not alone in their varying degrees of elation or apprehension. Judy Allen recalled:

> My parents, Stan and Ethel Dunstan had a dairy farm between Fish Point and the lake. Catalinas often made their approach over our farm. At first, our cows took-off with fright and tried to jump the fences. Dad and Mum were most alarmed, when for a period, the cows went off their milk until they became more accustomed to those huge, noisy monsters.

On the closing day in December, Squadron Leader Pratt arrived in Miles Hawk A37-4 to inspect the Depot's technical library, then conducted aptitude tests for airmen applying for commissions and recorded names of aircrew reservists.

As this momentous year drew to a close, the Lake Boga Depot strength had built to 618 personnel, including 21 Officers, 558 airmen, 1 ASO, 1 Nursing Sister and 37 WAAAFs. During the first five months of operation 23 flying boats had arrived at Lake Boga, of these 17 had required repair and service, with many being restored under the most primitive conditions.

With Japan's thrust south thwarted, Qantas would not have need to use Lake Boga; however, the volume of aircraft handled by No. 1 FBRD would increase significantly in the New Year.

An RAAF Vought Kingfisher floatplane on the ramp at Lake Boga in 1942. No. 1 FBRD did not repair any Kingfishers so the aircraft was likely on a training or communications flight from Rathmines. (AWM)

CHAPTER 2

1943

The 1943 New Year's Day holiday saw Lake Boga's public reserve crowded as adults and children picnicked under the willows or dived and swam from the wooden jetty.

For many civilians, this day provided the first opportunity on which to witness a flying boat alight upon the lake. As if to please the public, No. 11 Squadron's Flying Officer Bill Miller brought Catalina A24-14 in on final approach, its wing-tip floats lowered as the aircraft's shadow flicked past the limestone cliffs, then touched down just out from the reserve, disturbing the wild fowl as the flying boat's long wake rippled away in a giant V over the lake's glassy surface. This tranquil sight belied the turmoil through which this graceful craft and its crew had recently flown.

In common with all RAAF Catalinas, A24-14 had been hard worked, as had its crews. Prior to this aircraft's arrival, its second at Lake Boga, the Cat had been regularly flown, deep into enemy territory. Operating out of Cairns it had often flown night missions over Japanese airfields in New Guinea, as well as anti-submarine patrols over friendly convoys.

Arduous hours aloft, the stresses of constant take-off and alighting, and the mooring up in salted seas had taken its toll. Depot personnel proceeded with the myriad tasks associated with restoration as A24-25 underwent an intense seven-day overhaul.

Visits and postings to mid-month included an RAAF Officer from HQ Melbourne instructing in electroplating and an assistant Provost Marshal who met with Service Police. Air Commodore EC Wackett made his second inspection of the facility, while Chaplain E Bate arrived on a posting for "Methodist Chaplaincy duties".

The heavy drone of a multi-motored aircraft on the morning of 16 January turned eyes skyward as a giant Empire Flying Boat bearing RAAF roundels entered Lake Boga air space. No. 41 Squadron commander, Squadron Leader John Hampshire DFC flew a descending circuit and at 85 knots brought A18-13 down to a smooth alighting. His right hand was placed securely on the multi throttled quadrants, pulling back power as the flying boat's four Bristol Pegasus engines responded to his command. This Empire boat was in from Rose Bay, Sydney, and was internally fitted with frames for transportation of aircraft engines. Urgently needed Pratt and Whitney motors were then off loaded and gingerly floated to the Lake Boga public reserve where a temporary travelling gantry had been erected beside the jetty to facilitate the handling of aircraft engines and heavy materials.

This C Class Empire Flying Boat was the first of six such large twin deck, luxury aircraft delivered to Qantas Empire Airways for the Sydney to London service, but most often flown on the Sydney to Singapore leg, its peace time registration VH-ABB, and it was named *Coolangatta*. A18-13 was one of the C Class flying boats chartered by the RAAF for war service in 1940, when they and their crews joined No. 11 Squadron.

Within hours of arrival at Boga, the squadron leader prepared to depart for Rathmines. First the take-off check. T-M-P-F-F: Trim tabs, Mixture, Propellers, Fuel, Flaps. The check complete, fuel boost pump on, Flight Engineer Keith Zierk had engine air intakes set on cold, engine gills part open, fuel pressure pumped up. Switching ignition to his "outer port", Hampshire depressed the starter, his Flight Engineer the booster coil. As an outer engine fired, Zierk gave a final pump to fuel pressure while Hampshire moved his throttle setting to auto rich and adjusted propeller pitch. Again, the same procedure, now both outer engines were ticking over at 1,000 rpm, and warming. With the mooring line already slipped, alternate inner engines fired into life and the aircraft taxied downwind on the light breeze towards its take off point.

Wheeling around and steadying, Hampshire throttled his "outers" to take off boost, and brought his control column hard back into the pit of his stomach. In a wave of spray and gathering speed, the Empire's bow gradually rose, as the inner engines roared to 2,700 rpm. Easing his control column just aft of neutral, Hampshire, gained full rudder control with increased speed quickly throttled to correct a swing. Thundering across the lake at a full 85 knots, the flying boat captain steadily eased his great 36,000-pound (16 ton) craft off the water, clearing the distant willow trees with ease. His navigator then set a course for Rathmines.

On that particular Saturday, local residents relaxing along the foreshore reserve, had a fine view of the Empire Boat's arrival and departure, but would now wait over three years before a four engined flying boat would again grace the waters of Lake Boga.

RAAF Empire Flying Boat A18-13 seen moored in Port Moresby harbour in 1941. In January 1943 the same aircraft visited Lake Boga. (AWM)

From January, aircraft being received at Boga began displaying large letters on the waists of their fuselage, in addition to their RAAF registration. Instructions had been issued that month that each squadron's aircraft should bear identifying letters, with the paired letters denoting their squadron. Henceforth No. 11 Squadron aircraft bore the prefix "FJ" and No. 20, "RB". As aircraft came on charge with new RAAF flying boat squadrons, No. 41 Squadron was allotted "DQ", No. 42 "RK" and No. 43 "OX".

By 27 January, the second Depot slipway was complete. With experience gained during construction of the first concrete slipway, and with assistance from improved weather conditions and equipment, Neil Worner and his SRWSC team had

Lake Boga, an inviting alighting area. The Depot with waterborne aircraft is on the left, while the camp, an extension of the Lake Boga township, is on the right. In the foreground is Long Lake. (RAAF Historical)

completed this second slip in just ten weeks, against eighteen weeks taken for the first slipway. As SRWSC gangs cleared away their gear, Air Commodore J Murphy arrived on inspection. This experienced engineering officer from RAAF HQ Melbourne, No. 4 (Maintenance) Group, would continue to monitor the standard of activities at No. 1 FBRD throughout the war.

The workload was building. During January a total of eight Catalinas had arrived for service and restoration while one had departed. Postings saw the arrival of Fitter IIA LAC Ranal "Ron" Webster. Ron, an accomplished pianist and piano-accordionist, would soon establish a popular RAAF dance band. As personnel attended the month end movie, the title seemed appropriate, if not the film's content. Bud Abbott and Lou Costello featured in *Keep 'Em Flying*.

Having endured January, depot personnel were now coming to terms with the high summer heat. By night they sweated it out in the unlined barracks, and by day many endured great discomfort when working within the confines of intolerably hot aircraft fuselages. Then too there were the frequent dust storms.

The Duty Pilots Tower, with inflated windsock, near the lakes edge.

The summer dust storm phenomena had long been accepted by local residents as a normal, cyclical occurrence. Since the clearing of the Mallee for agricultural purposes in the late 1800s, the summer dust menace had increased. Early ploughing and harrowing techniques together with the movement of numbers of farm animals, fined out the fields and increased the wind-borne transportation of topsoil. Roly poly (tumble weed) became trapped against farm fences, facilitating rapid build-up of sand, with new fences built upon the protruding post tops of existing fence lines, a common sight.

While I do not wish to labour the dust problem encountered at No. 1 FBRD, it is of significance that of the vast number of ex-RAAF Depot personnel and aircrew contacted, the vast majority advised their main recollections of Lake Boga during war time were the dust storms.

Sam Roker, Draughting Section, who was posted to Boga in October 1943 told of a newly arrived airman just through his rookies. While he was standing guard duty, a blackout dust storm struck the Depot. The guard became so terrified and confused by this unfamiliar experience that he did the unforgivable, took cover in an adjacent building. When his mates heard of his action they convened a mock Court Martial, severely roasting the poor fellow.

On February 5, Corporal Sandy Sandow, a Fitter IIA working on a Dornier gun turret looked out from his hangar and thought "That Cat looks a bit different". Flying Officer Fleming then alighted in A24-29, a totally blisterless PBY-4 that had earlier served with the USN in the Philippines. This early model Catalina, in from Rathmines, was serving primarily as a workhorse, and arrived crammed with aircrew who would shortly take out flying boats A24-25, A24-30 and A24-34.

On February 7, Squadron Leader David Vernon DSO arrived aboard A24-2. During a 20-hour New Guinea mission on the night of January 6/7 he had released flares over an enemy convoy which revealed three escorts and five transport vessels. In deference to the assembled enemy fire power, a decision was taken to bomb at 5,000 feet. Tense minutes ticked by involving several aborted bombing runs, with Vernon's meticulous Flight Lieutenant Leslie finally shouting "bombs away". Squadron Leader Vernon:

> I immediately put the aircraft into a steep turn so that I could see the results of the attack. However, before I got around far enough there was a series of explosions and the aircraft received a real kick in the tail. As the ships came into view, I saw our target explode throughout its length.

Post war information confirmed the sinking of the 5,447-ton *Nichiru Maru* while transporting personnel of the Japanese 111/102 Regiment of which 456 men were killed and 85 wounded in this attack.

The skills of WAAAF personnel in all musterings were gaining recognition. WAAAF drivers in the Transport Section were among those acknowledged as a valuable element within the Depot's operation. The Transport Section's Flight Sergeant Colin Stewart:

> We always had several WAAAFs in our Transport Section ... They drove all of the RAAF vehicles, with the exception of the big flat top tenders. They drove them splendidly.

Expansion of Depot facilities continued apace as work programs continued. On the northwest corner of the Depot site a machine gun testing range was built, followed by excavation work for the receival of fuel tanks. Soon a 12,000-gallon tank for motor spirit was installed, together with two 12,000-gallon tanks for high octane aviation fuel.

Leo Allen mans a 0.50-inch machine gun in a Catalina's blister. Armour plate was not usual.

In the Lake Boga-Swan Hill region the grape harvest was now in full swing. At weekends, many personnel went grape picking, as much for the novelty as for the meagre remuneration. Sergeant Dick Clapson recalled:

> My mate Laurie Delaney and I used to pick grapes some weekends or sometimes go down the road to Lowe's who grew rock melons just for the seeds. We picked them by the truck load and fed the station - as long as we saved the seeds for Edgar.

During the month, inspections were made by RAAF Principal Medical Officer, Wing Commander Oxen, together with RAAF officers involved with administration, equipment and rehabilitation. By the end of February, the local night soil removalist's dreams of a quick fortune were dashed when he discovered planning was underway for a waterborne sewerage system to the camp.

While Fitters IIE handled aircraft engine overhaul at Lake Boga, major engine reconditioning was carried out at RAAF Station Tocumwal, NSW. Once removed the radial engines were fixed to steel frames, wheeled to the engine house or loaded aboard the Depot's flat top tenders, when Transport Section personnel drove the 186 miles to Tocumwal. Hay grazier Ted Dowling was stationed at Tocumwal and was directly engaged in testing the large numbers of aero engines that came off the station's reassembly benches. Ted Dowling:

> After joining up I was trained on overhauling, in line engines, Allisons and Merlins, then sent to Tocumwal to work on radial air-cooled engines. The RAAF had a huge setup at Tocumwal. After an engine was reassembled, it was wheeled to the test stand, as the big concrete building was known, the engine was secured, connected to appropriate instruments and each cylinder was fitted with a short, straight out, exhaust. After a four bladed wooden propeller or "club" was fitted and the thick doors closed and barred, we fired up the motor, running it at various speeds from 1,000 to 2,750 rpm with short bursts above those revs. The noise was unbelievable, the best place to be was in the control room, recording the engine's performance. We ran each engine for 8 hours. Particular attention was paid to the oil consumption.

Teams of nine personnel worked three shifts of eight hours, around the clock. The engine noise was so great that those involved were stood down every second week.

Over 2-4 March Allied aircraft scored a tremendous victory over a Japanese convoy known as the Battle of the Bismarck Sea. The victory was made possible because of Catalina A24-14 which had left Boga in early February. During the night of 2/3 March Flight Lieutenant Terry Duigan and his crew shadowed the convoy north of New Guinea and paved the way for the highly effective airstrikes the following morning.

On March 6 the RAAF's most senior officer, Air Vice Marshal George Jones arrived at No. 1 FBRD on an inspection visit. As AVM Jones cast a discerning eye over this expanding RAAF establishment, he was certainly in a position to evaluate all that he surveyed. A WWI fighter pilot with the Australian Flying Corps, after the war he had briefly flown a civil Avro 504K aircraft delivering the *Sunraysia Daily* newspaper to townships along the Murray River. However, this was short lived and in 1921 Jones joined the fledgling RAAF where he applied himself with great determination as he progressed within the service.

Fire broke out at the Depot's Sick Quarters 11 March with damage limited to the Sister's sleeping quarters, her sitting room and the patients' recreation room. Then another fire! At

The morning after! The sergeant's mess was destroyed by fire on 21/3/43. (Neil Worner)

approximately 1940 hours on Sunday 21 March, Depot HQ learned that the Sergeants' Mess was ablaze. Sheets of flame shot skywards attracting a large number of onlookers as the Depot firefighting squad struggled to contain the conflagration. The official report later noting:

> The firefighting squad was quickly on the scene but was unable to control the flames due to the poor pressure of water supply. The building was gutted in about half an hour.

RAAF procedure required that a court of enquiry be held into the cause of the fire. Accordingly, two days later, Squadron Leader J Toohey, commanding officer, Station HQ Tocumwal, together with three officers from HQ Melbourne arrived to conduct the enquiry. Geoff Marshall recalled that during the enquiry a sergeant who had arrived early on the scene of the fire was asked what had been the first action he had taken. With stem countenance the sergeant replied "I saved the beer Sir".

A PBY 5 Catalina bearing Dutch markings arrived on 25 March, the first of many non-RAAF Catalinas that would alight over the following three years. The flying Dutchmen had arrived at Boga! This was Catalina Y-45. In for an overdue engine change and repairs, this aircraft bore the recently revised Dutch markings of orange, white, and green horizontal bands, painted on the aircraft's fuselage and mainplane.

This aircraft had been serving with No. 321 (Dutch) Squadron in Ceylon. On 16 December 1942 Y-45 had departed Ceylon on the first nonstop flight to Australia, fitted with extra fuel tanks and carrying seven tons of fuel. It arrived at Exmouth Gulf, WA, 26 hours and 15 minutes later. After a brief visit to Perth the Catalina returned to Ceylon a few days later, before returning to Australia in March for an engine change at No. 1 FBRD.

The practice of involving ground staff in test flights continued, the theory being that if you had worked on a particular aircraft, then you should climb aboard and discover if the flying boat performed to the satisfaction of the musterings involved in its restoration. The number of personnel taken on individual test flights was increasing with the largest yet recorded on 24 March when Squadron Leader Geoff Marshall took off on test in A24-21 with a total of fourteen crew and Depot personnel. WAAAFs were often included.

Experience had taught the RAAF that where a unit is established, provision should also be made for deaths and subsequent burials of personnel. Accordingly, No. 1 FBRD's Operations Record Book reveals:

> March 26. Captain RS Sharp from the War Graves Commission visited the depot and inspected the Lake Boga Cemetery with a view of the possibilities of accommodating airmen's and service women's graves.

Within two weeks of this visit, the first fatality among Depot personnel occurred when LAC Ernest WM Hollebon, an electrical fitter, died as the result of a motorcycle accident at Koondrook near Kerang. A military funeral was accorded the deceased airman, after which his body was interred in the War Graves section of the Lake Boga Cemetery.

Towards month's end, great activity was seen out on the lake as a SRWSC team proceeded with the installation of the final group of mooring buoys. Two-ton flat peaked pyramid shaped blocks of concrete were manoeuvred out on the water and lowered on to the lakebed while stout cables secured more of the inverted, pear-shaped rubber buoys. Manufactured to stringent specifications by Dunlop Australia, thirteen buoys now lay out from the Depot area. From each buoy, four rubber covered cables protruded above the water line upon which flying boat mooring lines could be readily attached.

The two-ton concrete anchors had been poured at the lake's edge and an ingenious platform constructed to transfer these blocks out to their surveyed positions. Two substantial cylindrical steel tubes with sealed ends formed pontoons to which a platform with a 24-foot high, "A" framed jib was hoisted. The concrete blocks were individually craned out into the water, the pontoon moved over the block, the jib cable attached then transported out and lowered into position through the open centre of the pontoon.

In early March, an air and ground inspection of the Lake Boga facility was made by DOI officers, and perceived shortcomings reported upon their return to Melbourne. This report criticised the Sick Quarters, at four miles distant, as being too far from Lake Boga, the Emergency Aid post attached to the Dental Block, as unnecessary, and the underground communication, signals, and cypher building, far too substantial. With regard to efforts taken to have the RAAF living quarters appear an extension to the Lake Boga township the report stated:

SRWSC Pontoon and jib. Two-ton concrete buoy anchors were moved out under the pontoon and lowered into position.

> I am afraid that, from the camouflage point of view, it has not been very successful as from the air it is quite evident that the new roads and buildings which are devoid of any vegetation, is entirely something that has been recently added to the existing township.

A couple of cunning LACs had brought themselves undone. With installation of underground fuel tanks about to proceed Depot fuel was still stored in 44-gallon drums by slipway No. 1. Having access to an old Ford car but little petrol, one evening the two airmen rolled a 44-gallon drum of fuel into the lake,

Inverted mooring buoys with rubber covered securing cables undergo inspection at a Dunlop Australia factory. (Dunlop Australia)

the drum floating just under the surface. Manoeuvring it some distance through the shallows then securing it in a reed bed, there the drum remained until leave passes allowed a quick trip to Melbourne. Armed with four-gallon cans and a funnel, the airmen rolled the drum ashore only to discover they had mistakenly "procured" 44 gallons of oil.

During March, a total of six flying boats had arrived at No. 1 FBRD while eight had departed. Depot strength stood at 817 personnel.

On April Fools' Day, personnel rose to see a woman's girdle soundly attached to a lamp post outside the WAAAF quarters. An investigation on the same date into the loss of the Sergeant's Mess attributed the fire to the inflammable bitumen-based insulation. The report was critical of the poor water pressure from the elevated tanks and the total absence of a booster pump within the camp area. An order was subsequently received for removal from depot buildings, of bituminous based insulation. With no alternate replacement specified, great disquiet arose among personnel who saw what little comfort and protection they enjoyed from the extremes of the elements, evaporate.

While RAAF Catalinas continued with key operations attacking enemy targets in New Guinea, April would be remembered for the commencement of operations in the North West Area, and the inauguration of top secret mining operations flown by RAAF Catalinas. The Catalinas with their long range and high load capacity took up this challenging task. The first mission was on the night of April 21/22, when eight RAAF Catalinas took off from Cairns and successfully placing sixteen magnetic mines in Silver Sound, near Kavieng on New Ireland.

However, the month was a bad one for the Catalinas. After one was lost in an accident at Rathmines without casualties, another crash-landed at Cairns with the loss of all six crewmen aboard. Later came news of A24-41 which went missing with a crew of nine during an operation over the Gulf of Carpentaria.

To mid-April only three RAAF Catalinas arrived at No. 1 FBRD, with a like number departing after service and restoration. A rare bird alighted at the Depot on 15 April: Dornier X-24, the sole surviving Do24K still retaining Dutch markings and still flown by ex-NEI personnel. In November it would join the RAAF as A49-6.

By now the Depot had its own Post Office, separate from that in Boga but manned by civil postal staff. Postal services were operating smoothly, but for one disappointed LAC who kept up his daily vigil without success. Because identification by mail of RAAF bases was forbidden, letters to Lake Boga personnel were addressed thus - LAC W Smith, Group 428, Australia. One airman short of funds wrote home to his parents, concluding his letter "things are crook in Tullarook, please send 10 quid". The dutiful parents promptly placed money in an envelope, addressing it to Tullarook. The letter never arrived, or so the story goes.

Six months had passed since personnel had first moved into the camp. By now a short cut between Camp and Depot had been worn by RAAF and WAAAF cyclists. The unit strength at the close of April stood at 753 personnel, comprising 28 officers, 666 airmen and 59 WAAAFs.

Lake Boga on Friday 21 May saw another RAAF Catalina test flight. A total of 23 personnel embarked on a test of new Catalina A24-53. As Marine Section launches motored out to the flying boat, the rut tut tut of machine gun fire, coupled with thuds of point 50 calibre ammunition thumping into the earthen mound of the new machine testing range, echoed across the water. The machine gun's sound had ceased as Squadron Leader Marshall fired up the tight Pratt and Whitney engines taxiing A24-53 out from its mooring. Engines warmed, take off check complete, Marshall swung his craft into the breeze as the heavily loaded aircraft first floundered, then nosed up in the water gathering speed, the crowded aircraft up and away on a one hour, fifteen-minute test flight.

That same month a signal was received advising the promotion of Squadron Leader Marshall to the rank of Wing Commander. All work and no play? Indeed no! Within all services recreation was an important element in maintenance of morale. During May the RAAF played Victorian Rules football against Swan Hill. Corporal Bill Taylor captained the RAAF, Matt O'Halloran, Swan Hill. The scores were perhaps indicative of local enlistment, the RAAF kicking 14 goals 17 behinds to Swan Hill's 1 goal 2 points.

It was a fine sight; sturdy feather fetlocked Clydesdale's hard at work within the Depot area, back filling soil over the 92-foot length of the semi underground communications building. Surrounded by beached flying boats, these

The radio and operations room, in the underground communications building. The signals office is through the window. The building was manned 24 hours a day in three eight-hour shifts. (Bernard Fitzgerald)

proud animals, whose breed had for decades helped master the Mallee, now pulled laden tumble scoops disgorging soil over the length of the arched concrete structure. Collared tug chain and swindle-tree took the strain, as local farmers skilled in working their horse teams made short work of covering and camouflaging No. 1 FBRD's nerve centre.

Meanwhile SRWSC engineer Neil Worner and his men continued with their works program. While supervising concealment of the communications building, Worner monitored the erection of the three, 70-foot radio towers. Three similar towers were subsequently erected at the RAAF transmitting station just east of the Lake Boga township. Upon completion, the towers were strung with cables by RAAF personnel.

On 1 June winter had arrived with a vengeance when bad weather swept the Depot and living quarters. Flying Officer Daniel Bond hammered through the lake's foaming waves, lifting off in the new A24-54 bound for Cairns however deteriorating weather conditions forced a return to Lake Boga.

In early June, by diverse means members of No. 41 Squadron travelled to Lake Boga to ferry out Dorniers A49-3 and A49-5 up to their base at Townsville, some aircrew having travelled south by train, an arduous and time-consuming undertaking. The fact that these two Dorniers had arrived at Lake Boga nine months earlier is an indication of the difficulties encountered during their overhaul and restoration.

Almost a year had passed since Depot construction had begun. By June the new parachute folding facility was in place, an extension of the paint, dope and fabric store in building 17. RAAF regulations required that parachutes be regularly aired and refolded. A vented rectangular, parachute flue, 26 feet tall, was erected topped by a rope and pulley that enabled personnel to hoist parachutes, canopy first, up the flue where they hung until aired and dry. A 36-foot long, parachute folding table completed the facility. In a request for improved conditions for personnel CO Marshall advised HQ that the temperature in the parachute folding area had been just 26°F at 0800 hours.

June also saw the completion of the centrally situated draughting hut. The detention cell and guard house by the main gate was now complete while sandbags sixteen high stood stacked to either side of the entrance to the Communications Building. While minor construction would continue, the Depot was near fully developed. In an effort to reduce the dust problem, further water reticulation to planned areas of grass was approved. Amongst all this building activity and constant manoeuvring of beached flying boats, an existing vineyard had managed to survive.

Some things it seems never change. The tale is told of a guard who one night challenged an unidentified person within the Depot perimeter: "Halt! Who goes there?" the guard shouted, "Recite, Advance Australia Fair!", "I don't know it!" came the reply. "Pass Aussie" the guard retorted.

By mid-month an ICI representative visited advising on electro-plating. Pilot Sergeant Potter of No. 1 Communications Flight flew into Swan Hill with a camouflage team on board de Havilland Dragon Rapide A34-20, while a separate party inspected the Swan Hill aerodrome.

Now on its fourth visit to the lake, and in need of a further engine change and complete overhaul Catalina A24-17 alighted, ferried down from Cairns by No. 11 Squadron's Flight Lieutenant Bill Miller. Among a series of demanding operations recently flown was one in March, rescuing survivors of a USAAF B-26. The downed aircraft had been hit during a raid on Rabaul in May 1942. For ten long months its surviving crew had been sheltered by islanders, and now the RAAF would attempt a night-time rescue. On 24 March Squadron Leader Reg Burrage departed Port Moresby but soon noticed the outside access hatch to the anchor compartment up forward unsecured. Moments later, WAG Sergeant Ken Jones responded to his skipper's observation by emerging from the Catalina's bow gun turret. With crew members grasping a leg apiece and in buffeting winds, Jones deftly leaned down and secured the anchor hatch, and when wrenched back to safety said, "It was a long way down to the water."

After landing within Japanese occupied territory at 0245, the three surviving US flyers came aboard with a Coast Watcher. Captain Burrage then gunned A24-17 out through an improvised flare path. Prior to an 0800 hour alighting, a signal had been despatched "Coming home with the bacon". The reception committee at Port Moresby's pier was most formidable with war correspondents, journalists and a large gathering of Australian and US personnel.

For some weeks, when shopping at Lake Boga and Swan Hill, local residents had noticed numbers of farming families were accompanied by men in distinctive maroon coloured work clothes. These men were the district's first Italian prisoners of war.

Alan Fitzgerald in *The Italian Farming Soldiers* reveals their story. The initial arrival of Italian POWs in Australia occurred in May 1941 when they were interned behind barbed wire in camps in all Australian states except the Northern Territory. By year's end 4,396 prisoners had arrived from Egypt, and another 13,207 were progressively brought to Australia from prison camps in India.

The nearest Italian POWs to Lake Boga were in camps at Hay and Wakool, NSW, but were not considered a security risk to No. 1 FBRD. By May 1943 a joint decision by the Minister for the Army and the Directorate of Manpower was implemented whereby Italians were released from POW camps to live unguarded while working on farms. The men were nevertheless readily recognisable by their distinctive maroon clothing.

At this time 97 POW Control Centres were established throughout the country, and to regulate this operation each centre was responsible for a maximum of 200 prisoners. Within Victoria, sixteen such centres were established including those at Swan Hill and Kerang.

Modified and fully operational by 21 June, fitted with machine guns to bow, blisters and back hatch, bomb racks in place and pilot's armour plate removed, it was hoped that departing new Catalina A24-55 would meet with more success in the war zone than it had upon arrival in Australia. When on the Brisbane to Rathmines leg of its delivery to Lake Boga, the aircraft's captain, when off the NSW coast near Kempsey, caught sight of a large column of water. Nosing down to investigate, Pilot Officer Robert Honan and crew discovered wreckage and survivors of a just sunk Australian merchant vessel. Seconds later they spotted an enemy submarine at periscope depth. Pilot Officer Honan:

> Oh! The frustration was unbelievable. There we were without any means of attacking an enemy submarine which had just sunk a ship off the east coast of Australia. Our aircraft was brand new and not even fitted with bomb racks or gun mountings. I momentarily thought of giving the periscope a nudge with my wing tip or hull, but common sense ruled that out.

On June 26, the Chief Maintenance Officer, HQ, No. 4 (Maintenance) Group, Wing Commander Stevens had already arrived at the Depot from Tocumwal aboard RAAF Stinson A38-1, when Dutch Cat captain, Lieutanant Jedeloo alighted in Catalina Y-85. The aircraft received a modification and had its unsatisfactory self-sealing fuel cells removed before departing two days later for Fremantle "wet winged", this being the first recorded, westerly flight, Lake Boga to Perth.

June had proved a lean month with the arrival of only five flying boats and the departure of seven. The midyear strength numbered 755 personnel comprising 29 Officers, 668 Airmen and 58 WAAAFs.

At 64 Campbell Street Swan Hill, on 6 July, an All Services canteen opened its doors. This comfortable facility was established by the RAAF Women's Auxiliary headed by the Depot CO's wife, Joyce Marshall. At this same time the Junior Red Cross ran a Susan Competition as a fund raiser in support of Australian POWs, Susan being the Marshalls' daughter.

Two flying boats had arrived at Boga by 10 July. A Dornier from No. 41 Squadron, Townsville, and a Catalina, A24-38. A24-38 had been briefly attached to Sikorsky Kingfisher equipped No. 107 Squadron, RAAF, which was flying continual anti-submarine surveillance along the Australian southeast coast in response to persistent Japanese submarine activity.

During training Signals Section personnel were required to take an oath of secrecy, and as ACW Ethel Rowland (nee Pfeiffer), posted to Boga in 1944, recalled:

> The WAAAFs from Signals Section were required to sleep at the end of our hut and were not to openly discuss RAAF communications.

With this clamp on all information handled down in the Depot's communication building, WAAAF signals personnel on duty, quietly chatted among themselves:

> Yes, it was true, the signal has gone to the duty officer, the Yanks are coming!

On Sunday 11 July to those who glanced skyward the markings on the descending black Cat's bow and mainplane appeared distinctly different. A white five-pointed star flanked by two horizontal bars. Hundreds of miles from any ocean, the first of many United States Navy flying boats then alighted upon Lake Boga.

Depot administrative personnel were soon to discover the ranking of US Navy aviators: First pilot or Patrol Plane Commander – PPC, second pilot - PP1P, third pilot - PP2P and Flight Engineer, CAP - Chief Aviation Pilot. A Lieutenant (jg) was a Lieutenant (junior grade), one rank below a full lieutenant. When entering their flight log books, US aviators divided hours into six-minute intervals, thus 5.5 hours indicated 5 hours 30 minutes flying time. The Buno was the Bureau Number, the registered number of USN aircraft while aircraft side or tail numbers were prefixed "#".

The first USN Catalina to arrive was BuNo 08227 which had flown in for a 240-hour inspection, an overhaul, repaint and major bow repairs after damage sustained in New Guinea. After transport to shore by the Depot launch, the crew was welcomed by the RAAF Duty Officer and then taken to their accommodation at Barwicks' Royal Hotel, Swan Hill.

Lt (jg) Walter Hartley, USN, from VP-101.

A second USN Catalina arrived on 12 July, followed by the third two days later. This was pilot by Lieutenant (jg) Hartley from Tulsa, Oklahoma, who had experienced some difficulty in locating the secret southern RAAF repair depot:

> Brisbane to Lake Boga was a flight of something over 800 nautical miles with few distinguishing landmarks. There were no low frequency radio stations and our aircraft was not equipped with an automatic direction finder. Our direction finder did not indicate whether we were bearing to or from a signal.
>
> After seven hours we came upon a town with a river. About ten miles from the town we saw a lake covered in bird life but no human habitation. I thought if that was Lake Boga the camouflage was sure something. I flew back to the town and spotted a school yard full of children. I had a message written "Point to Lake Boga" and dropped it on a long red streamer. I made a 360 degree turn and expected to find the children forming a giant arrow. Instead, the kids threw rocks at us. I then flew down the main street at 50 feet hoping to read the town's name. All I read were beer and ale signs. Lake Boga was finally located after voice contact with Canberra.

Lieutennat (jg) Hartley had flown over Balranald. The three newly arrived USN crews belonged to Squadron VP-101, from Patrol Wing 10, which since early 1942 had been based on the tranquil waters of the Swan River in Perth. They had found their operations base so idyllic, they named it The Swan River Flying Club.

For many months the squadron flew regular patrols off the Western Australian coast covering the Western Approaches out of Geraldton and the Northern Approaches out of Shark Bay and Exmouth Gulf. By the end of June 1943, Patrol Wing 10 had been ordered to relocate to Port Moresby, the location from which these three newly arrived USN Catalinas had just flown.

Towards the end of July the three USN Catalinas undertook acceptance test flights before departing Lake Boga for Cairns. At Cairns a number of USN PBY commanders flew with RAAF captains on familiarisation missions. The RAAF squadrons had been flying regular strikes against the enemy in the New Guinea area since June 1942.

Such flights are recorded in the book *The Black Cats* by Captain Richard C Knott USN (Retired):

> They [RAAF crews] loaded ammo and flares on the catwalk so high that you could not move around. We were so heavy, we ran 3 or 4 miles to get off. Then we stayed at about 500 feet in a nose up attitude and remained constantly at 95 knots. When fuel burned off and speed increased, the skipper reached up and pulled the power back a bit. All hands flew in khaki shorts. We slept until nightfall when we got into the enemy area and then began looking for targets. I do not recall what we did that night but do recall we were empty and nose down when on the way back. Back at Cairns we had been in the air for 24 hours 10 minutes.

Donald Hartvig has written to me that "Those RAAF pilots I talk of were rugged, tough, guts like you could not believe and they flew the PBY like a fighter once they found the enemy." Of his few days at Boga and Swan Hill he wrote, "… being 22 or 23 years old, single, virile, we … dated every girl in sight."

Lieutenant (jg) Hartvig did not mention the beer bottles. From early days, as an added harassment to the Japanese, many RAAF Cat crews had flown out with a supply of empty beer bottles, a razor blade in each neck. Thrown from the blisters they sounded much like small bombs on their way down.

Aircraft departures that month were balanced by arrivals and included a RAAF Dornier and two new Cats fresh from the USA. At the Depot Sick Quarters, Sister E Beggs now settled into her new posting while a YMCA officer inspected Depot amenities.

In early August Consolidated Aircraft Corporation's Charles "Charlie" Morris flew into Boga from Forward Echelon, Brisbane, and became the Depot's resident PBY consultant. One evening as he moved through the dimly lit Depot, Charlie was challenged by a RAAF guard. "HALT! Who goes there, step forward and be recognised" the guard demanded. "Don't worry buddy" Morris replied, "I represent the Consolidated Aircraft Corporation of the United States of America." "If you damn well don't step forward and be recognised you'll soon be represented by a bloody wooden cross" the guard retorted.

A USN Catalina being serviced at Lake Boga, with the "star and bars" just visible on the forward fuselage. (AWM)

At this time Cats A24-56 and A24-64 were departing Lake Boga for No. 43 Squadron, a new general reconnaissance squadron that had, in June, relocated from Bowen to Karumba in the Gulf of Carpentaria. The Karumba base had been the site of a pre-war staging point for Qantas Empire flying boats on the Sydney to Singapore route.

On 4 August the tall moustached Scot Flight Lieutenant John "Jock" Tennant arrived on a posting from RAAF Station Nowra. An officer with a wealth of experience Tennant had been one of the originals with No. 10 Squadron and a crew member on a Catalina delivery flight in 1941. Flight Lieutenant Tennant had also served as an instructor in the first conversion course undertaken at Sea Plane Training Flight, Rathmines. A colleague of accomplished aviator Squadron Leader GU "Scotty" Allan, Tennant would soon welcome his compatriot to Lake Boga when Allan became commanding officer of No. 1 FBRD. The skirl of Jock Tennant's bag pipes was now heard floating across the Depot camp, however Jock's attempts to teach CO Marshall the art of puffing the pipes were to no avail, CO Marshall preferring the violin.

The ear shattering nerve rattling, gloriously spectacular low flying continued apace during test flights, and upon aircraft departures as RAAF, US Navy and Dutch flying boat captains and commanders "beat up", "shot up", "zoomed" and "flat hatted" farmhouses, local townships, the Depot and its living quarters. WAAAF cook Mary Paragreen (nee Jones):

> We WAAAFs always found the low flying so exciting, sometimes they even bombed our camp with rolls of toilet paper.

Again, this regular spectacle did not delight all those who were witness to the antics of these daring young men in their flying machines, as evidenced by an August letter in the *Guardian*. Headed "LOW FLYING", a terrified lady exclaimed:

> Sir – Can nothing be done to stop low flying and stunting over the town? This morning one of those tree hoppers flew over at about 200 miles an hour, narrowly missing some telephone wires and high trees, and terrifying hospital patients and nervous people. Will we wait for a tragedy to occur before taking action?

Adding poignancy to the lady's plea, this letter appeared next to a large advertisement promoting Frazer and Horn, undertakers.

Where else but at the local hospital were young pilots and aircrew to find such a concentration of young ladies to impress. On a same subject, an RAAF captain recalled his frequent return flights to Cairns on completion of all night operations.

> No matter how tired we were after night ops, if we felt we had enough fuel left we always shot up the Cairns hospital just to let the nurses know we were back home safely and would be calling on them.

Captain Paul Stevens, USN (Retired), has written to me:

> I do have very warm memories of Swan Hill, and Lake Boga. The people were most gracious and sought to make our visit there a very pleasant one. And I do hope the folks have forgiven us for "flat hatting" the towns upon our departure.

By now Americans began moving freely throughout the Depot and local communities. While some USN aviators journeyed to Melbourne awaiting restoration of their aircraft, most stayed and explored the local environs. Until this time, few among the populace had ever met an American. The following eighteen months would certainly change that circumstance. The Yankee accent was now heard at first hand and seemed more pronounced live, than on the movie screen.

I can still recall two US aviators looking at the priced fruit display in the Paragon Cafe window that summer:

> Hey Homer, look whart they karl raack melons darrn hererr. Canteloopees!

These newly arrived Americans called cars, automobiles, holidays, vacations, autumn, the fall, weatherboard houses, frame houses, petrol, gas, girls, gals and flying boats, seaplanes. There was no doubt the American officers looked impressive in their well-tailored uniforms and leather flying jackets, worn with gold badged caps. In those austere Australian days the Americans' new style silver framed sunglasses and zippered multi pocketed valises also attracted attention. Besides, they all appeared burdened with cartons of Lucky Strike, Camel and Chesterfield cigarettes.

Personnel frequently accepted hospitality on local farms. One airman recalled day leave with a farming family who collected him in their old Dodge car, then, in deference to petrol rationing, returned him to the camp at sunset in their horse drawn buggy. Depot Post Office telegraphist Ken Jones remembered visits to a local farm during which he travelled half the distance by bicycle, the balance in his host's motor car.

More than 50 years on it is difficult to convey the problems and discomfort associated with wartime road and rail travel, Tourist Bureau officer Kath Cramer (nee Thompson):

> I was employed in the Tourist Bureau to the side of the Royal Hotel in McCallum St. There was always a string of airmen wanting seats on buses, Ansett's to connect with the Adelaide train and Murray Valley Coaches to connect with Albury's, Sydney train. Airmen pleaded with me to find them a seat.

A Leave Pass was an essential element within service structure and leave passes were a prerequisite when moving out from the area. Officers absenting themselves from the unit also recorded particulars in the Officers' Warning Out Book stating their destination, duration of absence, and contact telephone number. This procedure proved efficient with the exception of one despondent officer who recorded his destination as: "walking the streets of Melbourne."

On September 6 the RAAF's last new waterborne only PBY-5, Catalina A24-68, was flown on an acceptance test by Wing Commander Marshall. The following day A24-68 cleared the lake bound for Karumba and receival by No. 43 Squadron.

When Dutch Commander AJ de Bruyn and crew departed the Depot on the 4 September, big Harry Gryzen was again seated securely in his accustomed position up in the flight engineer's tower of Dutch Catalina Y-87. During earlier visits to the area, Harry had become a friend of my parents and encouraged me to count in Dutch, advising:

> Now you practice your Dutch counting. If the Japs don't get me, I'll expect you to know one to ten by the time I get back again.

I had found first year French at the Swan Hill High School somewhat difficult, but more than fifty years on, I am still able to manage, een, twee, drie, vier, vijf … Shortly after this visit Gryzen departed to the USA for training with the Marines.

Personnel and visiting aircrew continued to enjoy the hospitality extended by local townsfolk. Before departing Boga, American Lieutenants Ranney, Maychek and Cheverton were entertained at the Splatt Street, Swan Hill residence of Bill and Marjorie Whitlock. At the conclusion of the evening, Ranney called from the gate, "We'll give you a buzz as we leave in the morning." True to his word on Saturday 18 September, Splatt Street's early morning peace was shattered by the clatter of Cat engines, blasting away at roof top level. In spectacular style, Ranney banked Cat 08255 until it appeared the aircraft's wing tip would surely strike the 100-foot-high concrete water tower. As suddenly as the Cat appeared, it was gone, on up to the USN base at Palm Island, then to operations out of Samarai in New Guinea.

During September, No. 1 FBRD's outgoing Commanding Officer, Wing Commander Geoffrey Marshall set matters in order, subsequent to the transfer of command of No. 1 FBRD to Wing Commander GU Allan. However, CO Marshall still made time to carry out ten test flights in aircraft including one Dutch and four RAAF Catalinas, Dornier A49-3 and Wirraway A20-219, the wing tip of which had been damaged on landing by a trainee pilot from an RAAF OTU.

"We'll give you a buzz as we leave in the morning." The US Navy's Howard Ranney and crew buzz Bill Whitlock's Splatt Street home in PBY 08255 as they depart on Saturday 18 September 1943. (A composite photo, Brett Freeman collection)

Approaching Boga on 29 September, Catalina A24-33 flew its descending circuit, alighting in precise style, then taxied towards the waiting launch and mooring buoy. The Cat's captain then signalled the flight engineer to switch fuel flow to idle cut-off, and the Pratt and Whitneys coughed and died. The new incoming commanding officer, Wing Commander George Urquhart Allan, AFC, then eased himself from his cockpit, and with the crew, deplaned and came ashore.

Having conducted the new commanding officer on an inspection of the facility, Wing Commander Geoff Marshall departed the area in his trusty Morris 850 Roadster, the vehicle labouring under the combined weights of himself, his family, and Flying Officer Gordon Myers. As the Morris tootled down the highway towards Melbourne, securely screwed to the vehicle's dashboard was a Beaufort's metal disk upon which was inscribed "WARNING! Do Not Exceed 395 MPH in a Dive!" Wing Commander Marshall proceeded on attachment to RAAF HQ Directorate of Repair and Maintenance to review a series of Beaufort crashes at OTU East Sale.

The incoming Depot CO Wing Commander Allan had arrived at Lake Boga from RAAF Station Rathmines, where he had been CO and a senior instructor of Seaplane Training Flight. RAAF Captains and crew held Wing Commander Allan in high regard for his all-round abilities and felt fortunate to have had the benefit of his cautious skilled tutelage.

This 43-year-old, balding, blue-eyed CO, with the broadest of accents, hailed originally from Lanarkshire, Scotland. Within days, Depot personnel had adopted the sobriquet "Scotty" when referring, in absentia, to their new CO.

October 1, saw a new month and a new commanding officer. Quickly adjusting to the more austere environment than that of Rathmines and its Macquarie Lakes, Scotty Allan commenced his command with a strength of 27 Officers, 686 Airmen and 62 WAAAFs.

In the interest of Depot discipline, fraternisation

Wing Commander GU Scotty Allan, AFC, typically attired in roll neck skivvy with fellow Scot and Engineering Officer Flight Lieutenant Jock Tennant. Both men were on the 1941 delivery flight of RAAF Catalina A24-1. (GU Allan)

was officially frowned upon, but sometimes rules were broken. A certain WAAAF found herself "carpeted" by the OC WAAAF, accused of "walking down the main street of Boga, hand in hand with a married man". Admitting her "guilt", the WAAAF pointed out that they were friends; that she was engaged. "There must be more to it than that" persisted the OC WAAAF, repeating this or similar remarks to each denial from the ACW. The WAAAF, remembering an incident in which the OC had been involved, decided to play her trump card. "Madam" she said "there was no more in it than being out on the lake at three o'clock in the morning." There was a brief silence then an almost inaudible "Dismissed."

On 19 October Lieutenant William "Bill" Lahodney Jr brought in a USN Black Cat from Perth for repairs after it had struck a buoy. He noted:

> We flew to Lake Boga for repairs. Boga at Swan Hill, is a small 1890 town in the middle of the desert, sitting on the banks of the Murray River. We landed there without mishap and were very courteously received by the RAAF. Scotty Allan commanding, veteran with 12,000 hours. Were directed around by red haired Sheila Patrick, RAAF [WAAAF] Section Leader [Officer], and then ushered into town to the White Swan Hotel and another one. Our hotel was managed by an old couple and both displayed the spirit of the old Midwest. No running water in the room, and a medieval shower with two settings, damn hot and damn cold.

Somewhat refreshed, the Americans were ready to explore Swan Hill's main street, and in so doing discovered that the annual Swan Hill District Hospital Ball was being held in the Town Hall on that particular evening.

> At 12:00 we were ready to do the town but when we met the land lady at the door she said, "You'll be back by 11.00 PM, of course". "We always close at 11:00". She was squared away, but we had to awaken her at 3:00 AM.

> Wandered around the virgin territory and into a restaurant where beauty was not lacking. There we learned that the Annual Dance was to be held in the Town Hall. We staggered to the Town Hall and found it jammed with all classes and models of people from four years to 104 years. There were two bands who alternated on swing time and Australian Folk Dancing, we stumbled through a few Aussie numbers, Waltzing Matilda, etc, then met some very charming young ladies, one in particular Peggy Wood seemed like an intelligent and sensible girl as her sister Betty, and a cute little red head.

> We danced then sandwiches at midnight, danced some more, wended our way home through the quaint country streets of Swan Hill. We walked around the block at least four times and I prayed for a car, or something. Peggy lived in a white house, surrounded by a large lawn, trees, flower beds, and a white picket fence which added to this 1900 scene. Would love to see Peggy again sometime.

Barely 24 hours had elapsed in the busy life of this American aviator when he departed Lake Boga aboard USN Catalina #45 bound for the Northern Area.

> From Swan Hill to Brisbane, Queensland, Australia, dinner in town, unsuccessful landing on two gorgeous Aussie nurses, then Palm Island. Gad!

At Lake Boga, the fauna was having a trying time. Black swans, gliding into the lake collided with power lines, often hanging aloft, dead, in stark silhouette. The freshwater tortoises fared little better during their spring migration between Lake Boga and Long Lake. Crossing the highway in their hundreds, many met their fate under the wheels of passing motor vehicles. US Navy Lieutenant (jg) Walter Hartley who last cleared Boga on 27 October in Cat #2 enquired of the writer:

> Do you still have those godamn turtles on your roads? I remember we often times had to pull up so we wouldn't kill them.

LAC's Ron Webster and Norm Dohnt had their own recollections of the lake's reptiles. Ron of his hangar's pets, one with a RAAF roundel painted on its shell, another with Flight Sergeant's stripes. Dohnt recalled the artrocious smell after the tortoise died hidden in hangar crevices. Baby tortoises were sometimes sold to visiting US aviators. The rabbits proved fair game, as Fitter IIA Bob Owens recalled:

> After pay day, it wasn't safe to move round the area of a summer's evening. That's when the chaps bought a

An RAAF Catalina on the ramp at Bowen in 1943, by which time many of the RAAF Catalinas were painted black to reflect their nocturnal roles over enemy territory. (AWM)

fresh supply of point 22 ammunition. They went after the rabbits, and fired at everything that moved and lots of things that didn't.

With Black Cats in profusion, WAAAFs from hut No. 26 followed the trend. Moyna Don (nee Sands), Beth Meldrum, Elsie Cumming and Gwen Jones were among hut mates who acquired Nick, a black cat of their own. Dances and balls were popular. Lake Boga, Tresca, Mystic Park, Fish Point, Benjeroop Ultima and Nyah, all conducting their share of social functions.

Twelve months had now passed since the WAAAF had arrived at Lake Boga. While postings transferred dedicated personnel, new arrivals continually replenished WAAAF musterings. At this time ASO Sheila Patrick departed, while ASO Douglas, from the School of Administration, arrived as Officer in Charge, No. 1 FBRD, WAAAF detachment.

A sense of great camaraderie had long developed among the WAAAF now facing a second summer in trying conditions. By this date the WAAAF's had named several of their basic huts. Paint Section personnel had provided the WAAAFs with stylish signs. Huts proudly bore titles of *Blossoms in the Dust, Seldom Inn* and *Paradise Lost*. Recalling her Boga days ACW Palma Kenzor (nee McCallum):

> Some weekends we used to play rounders with the Yanks, they called it baseball. Discipline was fair. After lights out, we had both regular, and spot checks just to make sure all was well. With 800 airmen and some of the visiting air crew just across the road, I suppose that was reasonable.

The WAAAFs were not alone in having their domiciles adorned. Draughtsman Sam Roker shared a hut with administrative personnel and in the traditional Aussie put down of those in sedentary musterings, an airman had affixed a sign reading, *Chateau de Shines*, shiney bums.

On Wednesday 8 November things were tense in the underground Signals Section. Had the Depot lost its first incoming flying boat? Well past its ETA, US Navy Cat #6, had gone missing. A garbled radio signal had been received two hours earlier, then nothing. The minutes ticked slowly by on the great clocks in the underground building, then, through the switchboard came a call from South Australia.

Out of Rose Bay, the USN Catalina had failed to locate Lake Boga and continued on its south-westerly course until

Martin Mariner flying boat A70-3 in Bowen in May 1944. In December 1943 this had passed through Lake Boga after its delivery flight from the US. (AWM)

reaching the Indian Ocean at Kingston SE, near Robe, some 260 miles over distance. Fortunately for the six-man crew, the sudden appearance of a flying boat at Kingston had alerted local members of the VAOC, the Volunteer Air Observers Corp, a nationwide body entrusted with reporting unexpected aircraft sightings. Cat #6 was low on fuel, had engine trouble, but made a successful ocean landing. Overnight accommodation was found for the Americans, and the following morning, the engine problem was rectified. The navigator was provided with a more detailed map, and with assistance from the VAOC the Cat was gassed. The aircraft then battered its way across the ocean, lifted and was once again on its way to Lake Boga.

On 13 November personnel were saddened by news of the death of one of their number, Corporal Charles J Buchholz, who had been killed in a motorcycle accident. Following a funeral service, the 23-year-old corporal, from Monto, Queensland, was buried in the War Grave Section of the Lake Boga Cemetery. In due course the RAAF Register of Deaths and Burials recorded the circumstance surrounding Buchholz' demise under the requisite category: Ground Accident.

Towards month's end, the silhouette of yet another type of flying boat entered the area, circling the lake on its descending legs and alighted after what appeared a deceptively slow approach. Among a flurry of wild fowl, Squadron Leader Sam Wood DFC, brought in A70-1, the first of twelve newly acquired RAAF Martin Mariner flying boats. Safely down after a long ferry flight from the US, the flying boat taxied towards the marine craft which directed the Mariner's skipper to a vacant buoy. With a clatter, the Mariner's motors coughed and died, its massive four bladed propellers stilled. A large bow door swung open as a crew member with a boat hook secured a mooring line.

Through November, Mariners continued to arrive, and after beaching gear was attached, were unceremoniously towed, tail first, up the concrete slipway, then hauled to a hangar or placed in the open awaiting their turn for operational fitment, a coat of matt jungle green paint and service. The heavier Mariner beaching gear differed from that of Catalinas. Above each main wheel, which was part water filled, a large watertight duralium box was fitted to the beaching arm, thus providing balanced flotation during attachment.

November had been another eventful month. Four aircraft had departed, while fourteen flying boats had arrived: Catalinas, Dorniers and now Martin Mariners.

By this date a compulsive escapist, Italian POW Lieutenant Simoni had made a break from the Hay detention barracks and while being pursued, passed through the Depot's district. This escapee had authorities following his trail on horseback, Aboriginal tracker, boat and motor vehicle. Simoni rowed, swam and walked, leading his pursuers a merry chase via Balranald, on the Murrumbidgee River, Bannerton on the Murray River, then moving via Manangatang, Ultima and Boort to Bendigo where he managed to board a train heading south. Simoni's ten weeks of freedom would come to an abrupt end with his arrest in the Flagstaff Gardens, Melbourne, on 19 January 1944.

RAAF boxing tournaments in the camp's Recreation Hall continued to attract good crowds with the fight night on 29 November no exception. The main event was a vigorous bout of three, two-minute rounds between LAC Albert Goddard and AC1 Norman Moxham. The bell rang at ringside, and the fight was on. During the first round, Moxham delivered a blow that bloodied Goddard's nose, and then in the second round Goddard received a savage blow to the side of the nose and left eye. His nose bled profusely. Stopping the fight, the referee attended Goddard who advised the referee he was "all right". The fight proceeded with Moxham being declared the winner on points. Examining Goddard after the fight, RAAF MO Flight Lieutenant Rodger Edward considered the airman normal, but for a slightly injured nose. Suffering concussion the following day, Goddard was admitted to the Depot Sick Quarters from where he was transferred on 9 December to No. 6 RAAF Hospital Heidelberg, Melbourne. LAC Albert Goddard died there on Saturday, 11 December.

Friday 17 December saw the arrival of the RAAF's first PBY-5A amphibious Catalina, when Warrant Officer Cliff Hull alighted from San Diego in 34055. The aircraft was soon to bear its RAAF nomenclature, A24-69. A week later a flight of five Catalinas, of which three were new amphibians, alighted at Boga out of Rathmines. The arrival of these three new amphibious Catalinas was but a forerunner of 46 such amphibians that would arrive at Lake Boga over the next fifteen months.

The new Consolidated Aircraft Corporation's amphibious Catalinas were designated PBY-5A, the "A" for amphibian. Shortly after their arrival they received their RAAF serial numbers A24-70, A24-71, and A24-72. This versatile variation of the Catalina still bore the same classic configuration of its PBY-5 predecessors but had two large wheels that retracted into wheel wells situated in the sides of the fuselage, between the wing support struts. While the main wheels remained visible when retracted, the tricycle support system included a smaller bow wheel, forward, housed in a watertight compartment beneath the hull. When retracted, this wheel became enclosed by hinged panels in contour with the bow profile.

Squadron Leader J Radford, a medical officer from HQ Laverton arrived by air 21 December to head the Court of Enquiry into the death of LAC Goddard. This enquiry was additional to a civil Coronial Enquiry conducted on the evening of Saturday 11 December when the coroner found Goddard had died from: "injuries being received accidentally as a result of a boxing bout."

Christmas dinner 1943 was again served in traditional service style as Depot CO, Wing Commander Allan his officers and sergeants waited tables on other ranks. As 1943 drew to a close, many of the 821 personnel at No. 1 FBRD were bound for New Years Eve celebrations, while those of long standing looked back on a year during which the facility had become fully operational with a significant amount of work undertaken for the Air Forces of three nations.

Personnel of No. 1 FBRD on the massive wing of a Catalina in one of Lake Boga's hangars. (AWM)

CHAPTER 3

1944

The new year would see a dramatic increase in the Depot's workload with a forty percent increase in air traffic and a thirty percent increase in flying boat service and repairs. Conversion training to Martin Mariners had begun at Boga during the closing days of 1943 with No. 41 Squadron personnel and this program was now intensified. Squadron Leader Sam Wood headed an experienced group of Mariner captains providing daily instruction on the idiosyncrasies of the large flying boats. Such was the size of the new aircraft, pilots were required to mount steps in order to reach the bridge, or cockpit. Squadron Leader Wood:

> During conversion flights I preferred to fly in the mornings or evenings as I considered it quite dangerous while training to take off and alight after the summer heat had built up. While all looked calm, there was an enormous difference in air temperature between that over the dry land and the air over the lake's surface causing turbulence.

The twin-engine Mariner had interesting features. While having much of the bulk of Sunderland boats, it possessed a stylish gull shaped coarse dihedral inner mainplane. This swept up inboard wing section positioned the engines well clear of the water while the tailplane configuration consisted of twin rudders mounted on dihedralled spars. The entire trailing edge of the mainplane was hinged, the outer section acting as ailerons, the inner section as flaps. Mariners possessed a wingspan of 118 feet, a length of 77 feet 2 inches and stood 17 feet 6 inches high. Powered by two 14-cylinder 1,700 hp Wright Cyclone R-2600-12, twin row radial air-cooled engines, the aircraft was equipped with four bladed Curtiss Electric, constant speed, full feathering air screws. While many US Navy Mariners were armed, the RAAF flying boats arrived unarmed, readied for minor modification to transport aircraft.

In mid-January Squadron Leader Bryan Monkton ferried in Mariner A70-10. His passenger on this flight was Depot CO Wing Commander Scotty Allan returning on leave. The following morning Monkton was up early and on arrival at the depot heard gunshots from the nearby shore. Upon investigating, Monkton discovered his friend Scotty, with a double-barrelled shot gun crooked on his arm, eyeing off the water birds. Scotty Allan had been at pains to prevent the birds alighting on his moored aircraft and in the process, redesigning the paint work. Squadron Leader Monkton:

> Even with Scotty's ingenuity he could not prevent shags and other water birds alighting and defacing his aircraft.

Rainfall during 1943 had been only half the district average of twelve inches and the new year's figure would fall further to the second lowest recorded. Dust storms became more frequent.

With eleven motor vehicles now allotted to Transport Section, personnel were kept busy with a variety of tasks. A tale was now circulating concerning a WAAAF, newly posted to Transport Section, who thought that bottom gear was an item of female intimate apparel; the story was not taken seriously. Among other duties, WAAAF drivers delivered quantities of milk from the camp to the Paint Section. Milk was considered essential to the health of personnel working constantly within the vapor laden environs of the dope and paint shop. Construction of 28-foot-high windbreaks between selected hangars now proceeded giving protection to and dampening the movement of beached aircraft during storms.

Eighteen flying boats arrived at Lake Boga in January including five US Navy Cats, while three departed. With increasing numbers of US Navy personnel passing through the area, Ensign Herbert Franks was appointed US Navy Liaison Officer. As January closed, construction was completed by the foreshore, of a magazine to house tracer and ball ammunition, flame floats, aluminium sea markers and incendiaries.

An RAAF Catalina at Cairns in early 1944, with mines slung under each wing. (AWM)

The first day of February saw operational fitting and testing complete on the first two RAAF Catatlina amphibians and they departed for receival by No. 11 Squadron. Test landings at the Swan Hill 'drome had caused great interest.

With the arrival on 2 February of Mariner A70-12, all twelve RAAF Martin Mariners had now arrived in Australia.

After an earlier Depot inspection, a USAAF PBY arrived 5 February, the first US Army Catalina to be serviced. The USAAF classified Catalinas as OA-10 aircraft, and this newly arrived Cat was serial number 43-33261. By mid-month, another USAAF Fifth Air Force Catalina (41-118733) arrived from New Guinea. In a departure from accustomed USN black Cats, these aircraft had sea blue/sea grey camouflage for daylight air sea rescue work, or "Dumbo" flights as the Americans called them.

That same day, two USN VPB-33 Cats (#93 and #81) arrived, landing five minutes apart. PBY #8I was commanded by 24-year-old Lieutenant (jg) Robert Gates of Duluth, Minnesota, and had recently sustained damage from enemy gun fire. The aircraft had staged down from the Admiralty Islands for much needed repairs. Lieutenant (jg) Gates:

> Although I was a (jg), junior grade, lieutenant I was made senior officer in charge of six aircraft on an operation to Woleai in the Caroline Islands. Intelligence indicated that there were enemy ships in the harbor. After briefing the crews, we took off and flew into the night relying on celestial navigation. My PBY was flying at 1,000 feet with the others, split by five hundred. Sure enough there they were, a couple of ships there in the harbour. I pointed the nose down and got two of my bombs away, a 1,000 and a 500 pounder, and we had good results, (damage to both ships confirmed) except that we got a certain amount of ack-ack fire.

This operation was one of the first Allied raids on the Caroline Islands, and Lieutenant Gates' reference to enemy fire is perhaps a little low key. Damaged, and difficult to fly, the Cat's crew plugged the holed hull with fids, threaded wooden rods of varying sizes. Upon arrival at the base it was run up on the beach to avoid it sinking. Lieutenant (jg) Robert Gates:

> I was privileged to take the aircraft down to Lake Boga for repair. We subsequently flew to Brisbane, fuelled up there, then to Sydney, and pumped up on gas there. The next day we flew under the Sydney bridge which was kinda biga to me. Anyhow we motored down to Lake Boga, sure enough there it was just like advertised. We landed and the aircraft was taken into the environs of the facility. I and the rest of the crew went to an hotel in Swan Hill. Everyone was very friendly, we ate lots of steak and eggs. The beer was scarce, pubs opened at about four o'clock in the afternoon and we all got in line with the rest of the local beer guzzlers and got our share. There was a nice city park in which we played rounds of gin rummy. There or four days later some other airplane had finished its repair work and we got in that and drove back up to the war.

February's weather did not disappoint. Mid-month brought dust and a vast movement of Mallee soils, as noted in the Operations Record Book:

> A terrific dust storm with high wind velocity enveloped the depot late in the afternoon, necessitating doubling the guard and frequent inspection of aircraft moorings.

In a welcome break from duty, on the evening of Saturday 4 March, personnel again attended the RAAF swimming carnival at the Swan Hill pool. The evening was a great success; however, the mock Catalina personnel had built in which to perform an air sea rescue, sank to the bottom of the pool.

In mid-February, as a Country Roads Board team arrived to seal Depot carriageways, four more new RAAF PBY amphibians alighted on completion of their ferry flights from the US.

In March, the WAAAF celebrated their first official birthday, and the following day resumed their varied duties as a valued element within the Depot's structure. A large crowd attended a combined No. 1 FBRD sports day held at the parched Lake Boga football ground on Saturday 25 March. In merry mood, sand shoed personnel marched on to the ground padding past CO Allan and assembled officers.

Inter hangar rivalry was the order of the day. The Fabric and Paint Sections had been hard at work producing banners that team leaders now held high. A banner bearing a giant Busy Bee wielding portable machine tools heralded the arrival of teams from hangars one to three. Contestants from hangars four to six marched behind a much-muscled Popeye the Sailor Man astride a Martin Mariner. Popeye was pictured deftly dealing with the Busy Bee, and a Boxing Kangaroo, the latter the mascot of hangars seven and eight.

That evening, a minor battle erupted in front of Boga's Commercial Hotel. This skirmish involved civilians, the Army, the Navy and the Air Force. After an RAAF Sergeant attempted to drag an unruly civilian off to the local Police Station, the civilian's brother, a member of the AIF, took exception to the airman's action and an altercation ensued. The rumpus was joined by two "sailors" (USN aviators), another Boga airman and a civilian. The magistrate presiding at a subsequent civil court hearing decided that no action be taken regarding this disturbance.

At Boga an entry in the No. 1 FBRD Operations Record Book on 28 March recorded the crash of a Dutch Kittyhawk:

> Kittyhawk C3-524 crashed 2015 hours approximately, two miles southeast of Swan Hill aerodrome. The pilot was safe after baling out. The fire tender was despatched, and guards placed over the wrecked aircraft.

As it transpired, the pilot of this No. 120 Squadron Dutch P-40 fighter was certainly safe, but not completely sound. Twenty-six-year-old, Javanese born Dutchman, Second Lieutenant Arjen Johan Greets, had lost his bearings while flying across the sprawling Mallee wheat lands, but worse, he was running low on fuel. With his options shot, Greets pulled power on his twelve cylinder engine back to 1,500 rpm and then thrust back his canopy, wrenching himself up and out of the cockpit. The blast of a 160-knot wind hit him as he flung himself into the clear night sky. As Greets descended beneath his whispering parachute, his Kittyhawk spiralled down to destruction, exploding on impact, with rounds of 0.50 calibre ammunition providing an intermittent fireworks display. Fizzing tracer bullets, blown from their casings, illuminated roadside foliage as around the crash site fire spread through two acres of dry stubble. In an adjacent paddock Greets hit the ground with a thump, breaking an ankle, and was later transferred to the Swan Hill District Hospital.

Daylight disclosed the aircraft's twisted, burnt-out wreck, with its engine deeply imbedded in grain grower Keith Parsons' Lalbert Road paddock, debris littering the immediate crash site. By late afternoon, as No. 1 FBRD personnel transferred the last of the wreckage to the Depot salvage area, boys began arriving on bicycles, intent on inspecting the site and searching for souvenirs.

Whoops of excitement rose as the young fellows scratched through the blackened soil, unearthing various treasures. Diggers were well rewarded as shining exploded and unexploded 0.50 calibre brass shells were retreived. I managed to unearth the remains of a burnt canvas bag, from which segments of a shirt and shaving gear were salvaged. The razor blades had a khaki finish, while the burnt shirt's surviving epaulet bore a badge with a lion rampant, resting upon the word "Nederland". By good fortune, the shirt's breast pocket still displayed the pilot's bronze wings, beneath which hung a wreath, encircling the letter "W". I wondered whether this pilot had been decorated by his Dutch Queen, Wilhelmena. The wings were later found to denote a Vlieger-Waarnemer (Pilot-Navigator). In the circumstances, perhaps a little inappropriate (!).

A second Dutch Kittyhawk, C3-572, flown by Piolt Sergeant JD Brameyer met a similar fate on that same evening, the aircraft crashing some 50 miles from Mildura. The pilot parachuted to safety. The Dutch fighters had been returning to their Canberra base from a major alert at Exmouth Gulf in Western Australia.

The Dutch were having a bad time. As No. 1 FBRD salvage vehicles departed the wheat field with Kittyhawk wreckage, Swan Hill aerodrome was the scene for a mishap involving a Dutch B-25 Mitchell medium bomber.

With dust and gravel thrown skywards on Wednesday 29 March, the B-25 gouged its way down the limestoned airstrip. The nose wheel to the aircraft's tricycle undercarriage had collapsed on landing, with rising debris seen from the distant highway. Damage was sufficient that 0.50 calibre shells, spewed from the aircraft, dislodged from smashed, nose mounted ammunition chutes. Boys on bikes returning from the Kittyhawk crash site descended on the scene in the hope of acquiring further souvenirs.

One of at least two B25 Mitchells to land at Swan Hill, N5-182 had flown in from Canberra, and belonged to the Dutch No. 18 Squadron, based at Batchelor, in the Northern Territory.

On 5 April three Catalinas, a Mariner and a Dornier departed Lake Boga. First off the water was a USN Catalina bound for Brisbane, Palm Island and New Guinea. This was closely followed by Flight Lieutenant "Chestie" Bond who gunned his new Catalina A24-73 across the lake. It was being ferried out to No. 20 Squadron. Six weeks later this flying boat was lost during a mining operation on the heavily fortified port of Soerabaya. It was believed to have been shot down by anti-aircraft fire.

The annual Anzac Day march and commemoration service was held at Lake Boga with CO Allan taking the salute, after which RAAF Padres Millard and Rundle conducted a service. At Swan Hill an impressive Anzac Day parade was also in progress, when servicemen from three wars marched to the Cenotaph for the wreath laying ceremony, then joined local citizens in the Town Hall where the service continued.

During the month, on arrival at the Swan Hill District Hospital, an over-anxious RAAF ambulance driver had side swiped the wrought iron Garden Memorial Gates. CO Allan quickly arranged for the damaged pillar to be restored to operational serviceability.

With dances popular, April had ushered in the district's ball season with Ron Webster's Band providing much of the music. At a Boga dance one evening Ethel Dunstan (nee Woosman) of Goodnight, NSW was dancing with a US aviator when in the course of conversation he asked "And where do you come from Ethel?" "Goodnight" Ethel replied. "Hey mam" the American said "it's a bit early for that isn't it?"

While the profile of Catalina aircraft remained quite similar, the constant flight testing of Catalinas from three Allied nations required mental agility. Most fuel gauges in RAAF Catalinas were calibrated in Imperial gallons, US PBYs used US gallons while some Dutch fuel gauges were read in litres. In May, test flights continued with Scotty Allan at the controls. During a pre take off check, Scotty called to acting flight engineer, ex-No. 10 Squadron's Sergeant Bert Knight, for a fuel reading. "Aye, Sergeant, und hoe mooch fuel doo ye harve" crackled Scotty over the headphones. "Three hundred- and twenty-gallons skipper" Knight replied. "Aye. Am they Imperrrial gullons orr Arrmerican gullons, orr what?" enquired Scotty. "They're Australian bloody gallons" came the reply.

A Dutch B-25 from No. 18 Squadron, similar to that which crashed at Swan Hill on 29 March 1944. (AWM)

Late at night on 12 May Squadron Leader Reg Marr, with several crew members arrived by bus at Swan Hill to a dimly lit Campbell Street. Moving along the thoroughfare towards the Royal Hotel, Reg was startled when out of a darkened doorway a female voice said "Hello Reg". Squadron Leader Marr:

> To my great surprise it was my wife who had journeyed down from Sydney. She was known to do a bit of border hopping during those years.

The following day, Squadron Leader Marr departed Lake Boga aboard A24-91 staging to Darwin. A month earlier this No. 43 Squadron Catalina had been sent to rescue a Beaufighter crew whose aircraft had been downed during an attack on Timor. After a Mayday call, it was critical to make the rescue during daylight, so the Catalina departed without any surplus fuel. Two Beaufighters flew as escorts, and directed Marr to Cartier Reef where the downed airmen were sighted in a dinghy. This story is continued by Flying Officer Dick Udy:

> Now for the tricky bit, getting down on the water with them. Reg came in upwind, judged the swell and the position perfectly, pulled off the engines, a second or two of eerie silence, then CRASH . . . Full stall landings in a heaving sea were always dramatic, but we were down on the rolling waves tossing about.

The retrieval proved time consuming and frustrating but finally the two airmen were aboard. A take-off was achieved through a pounding swell and a head wind further depleted fuel on the 500-mile homeward journey to a blacked-out Darwin. The navigation was precise. With fuel meters on nil, the crew peered into the darkness looking for a flickering flare path. There it was. Coming in on direct descent, an engine cut out, the second coughed and continued. Safely down, the remaining engine sucked in its last sip of fuel and was silent. Catalina A24-44 was towed to its mooring.

Having arrived back at Rathmines from the US, Flight Lieutenant Harrison, Flight Lieutenant Foskett and Pilot Officer Honan exercised their new amphibian's wheels on 8 May, with a landing at Mascot Airport, Sydney, before proceeding to Boga. Soon after arrival these aircraft bore new RAAF registrations A24-96, A24-97 and A24-98. The Catalina captains and crews had experienced an eventful return journey to the US. Flight Lieutenant Ron Foskett:

> Australian crews who ferried Catalinas from San Diego to Rathmines and Lake Boga found that the rather opulent Officers' Clubs at the US Naval Air Stations had 50c poker machines which worked well with

Australian pennies. Needless to say a fair amount of copper currency went off shore in Aussie pockets to cash in on this bonanza. Something like a rugby scrum would form up around the machines at Kaneohe, the NAS on the opposite side of Oahu from Honolulu, pulling US 50c coins and carefully leaving one or two in the feed windows. The crews would then be quietly drinking the proceeds in a far corner of the club when an outraged Yanky voice would scream "Those - Aussie sons o bitches!" as they contemplated a handful of worthless pennies.

As the three RAAF ferry crews were coming ashore, battered Cat A24-35 was towed to a vacant hangar from its overnight position on slipway No. 1. This Catalina had literally "been in the wars" having weathered collisions at its moorings on three occasions, while its hull had twice been severely damaged by enemy action. Two weeks earlier the flying boat had been one of six from No. 43 Squadron that had mined shipping lanes near Balikpapan in the face of heavy anti-aircraft fire. An exploding shell blew a three feet square hole in the hull of A24-35, which the crew plugged up with bedding and parachutes. On arrival at Yampi Sound in Western Australia the pilot executed a nose up landing, before skilfully running the flying boat onto a narrow beach. Temporary repairs were affected, after which the aircraft was flown to Lake Boga.

In June, three weeks had passed since new RAAF Catalina amphibian A24-102 had arrived at Lake Boga. Fitment and modifications were complete, Scotty Allan had test flown the aircraft, and it now rested ready on the water. For some time, the engineering section had been concerned by the time consumed in removing and reinstalling flying boat engines. Now Scotty had an idea. Sunday morning 10 June at 0814, and in Depot crewed A24-102 Wing Commander Allan was away sweeping through the morning's mist and lifting, then setting an east, south easterly course. One hour ten minutes later his Cat's wheels touched the tarmac at RAAF Base Tocumwal. Wartime needs spawn wartime deeds, and Scotty was about to vary the RAAF' s interpretation of "procurement".

With the acquiescence of the Tocumwal CO, Scotty "borrowed" three long lengths of heavy steel beam. LAC Jack Settle, a crew member involved in that day's clandestine mission, recalled:

> The boys crewed for Scotty. At Tocumwal we soon appeared from the stores area with a steel beam over 30 feet long. This beam was gingerly manoeuvred into the waiting Catalina, up through the rear gun turret hatch, the beam's leading end passing right on through the pilot's compartment, under the control saddle, finally resting within the forward bomb-aimer's space. This procedure was twice more repeated. When the time came to depart, the three beams took up so much space that some crew re-entered the aircraft through the cockpit roof hatch.

By 1645 hours that evening A24-102 had alighted back at Boga. Approaching the slipway, drogues were cast out as Scotty lowered the Cat's wheels using them as water brakes. Then as his wheels touched the concrete slipway, he drew up the wing tip floats. The Catalina was towed to building 66 where the illicit cargo was disgorged in preparation for its new application. Wing Commander Allan was soon to admire his substantial new tripod gangtry. In future, engine changes would become more efficient!

Activity was brisk in June with nineteen RAAF and USN air movements to mid-month. Late on the 16th, out on the slipway, personnel wrestled waist deep in freezing water attaching beaching gear to A24-6 l. This aircraft was fortunate to have survived a May mining mission when its port engine failed while flying low over Soerabaya Harbour. The aircraft stalled and dropped, and with the increased load the starboard engine faltered then fired again. A mine was dropped to lighten off, the port engine spluttered back to quarter power. The pilot wheeled around and again swept in, determined to lay his starboard mine. Adrenalin pumping, the task was accomplished. Homeward bound, gear was jettisoned and the problem port motor was cut. The eight-hour flight home was safely accomplished on one engine.

Many Catalina crew members owe their lives to the reliability of a sole fourteen-cylinder Pratt and Whitney engine droning on hour after hour when the other was out. After a 10-hour 30-minute flight to Cairns on one engine pilot Damien Miller famously said:

> I've got two great friends in this world, Mr Pratt and Mr Whitney!

June 25 saw the end of a flying boat era when Flying Officer Dudley Wright alighted in Dornier A49-2. All five surviving ex-NEI Dornier Do24Ks were now at Boga where they were withdrawn from service and stored in a holding area. Availability of spare parts had long plagued maintenance staff, and the Dornier's slender fuselage had a restricted load capacity. The ageing aircraft had become increasingly unreliable and were unpopular with pilots, particularly after the arrival of Martin Mariners and newer model Catalinas.

On June 29, newly painted matt black Catalina A24-100 departed Lake Boga for No. 11 Squadron. The engineer on the first leg of this flight was Ivan Clempson, who recalled years later:

> I usually flew on operations with Flight Lieutenant Jack Hill and his crew. Today I have trouble remembering my car's registration number; however, I can still clearly recall the firing order of those twin row Pratt and Whitney's: 1-10-5-14-9-4-13-8-3-12-7-2-11-6.

A third airman died as the result of a motorcycle accident while returning on leave, again in the Kerang district. Melbourne born LAC Edward Morris Hircock died in the Kerang Bush Nursing Hospital on 29 June, following an accident on the previous afternoon. Hircock was buried in the War Graves section of the Lake Boga cemetery.

To mid-1944 almost 200 flying boat movements had been recorded. More than 63 aircraft had been serviced and restored. Unit strength on 30 June stood at 29 officers, 788 Airmen and 77 WAAAFs: a total of 894 personnel.

RAAF flying boat operational bases were undergoing restructure. Early in June Nos. 11 and 20 Squadrons had flown 60 personnel from Brisbane to Melville Island to help establish the home of newly formed No. 42 Squadron, the fourth and last of the RAAF Catalina squadrons. By late June, No. 11 Squadron received orders to relocate from Cairns to Rathmines while No. 20 Squadron would soon move to Darwin. No. 11's instruction was not well received, as David Vincent recorded in *Catalina Chronicle*:

> Thus the two famous Catalina Squadrons Nos. 11 and 20 were finally separated after serving together for almost four years through the hardest years of the war in the Pacific. No. 11 Squadron particularly, found this order hard to believe.

As No. 20 Squadron personnel crammed their Catalinas with gear for transfer to Darwin, the Martin Mariner equipped No.41 Squadron arrived from Townsville replacing the departing Cats at Cairns.

On 5 July the Depot's underground switch board received a phone call from Bill Laird, the manager of Murray Downs merino stud and sheep station in NSW. A Wirraway aircraft from No. 2 OTU at Mildura, had landed in the paddock opposite Murray Downs' main entrance. Low on fuel, a trainee pilot had come down on what appeared an ideal landing area, an area that until 1937 had been the site of Swan Hill's original aerodrome. As the pilot touched down, he was unaware that his Wirraway's wheels were retracing the path of the famous Fokker tri-motor aircraft *Southern Cross* and the path of Australia's most famous aviator, Sir Charles Kingsford Smith. Almost eleven years earlier, in November 1933, Kingsford Smith had spent two days at Swan Hill during a barnstorming tour of south-eastern Australia. A substantial crowd had turned out to see Smithy and his renowned *Southern Cross*, many people boarding the aircraft for a ten-shilling flight.

During July, ten USN Black Cats arrived at Boga. Depot personnel were soon at work on these growing numbers of USN aircraft. Away from the war the Americans often strolled through Swan Hill, sometimes playing penny on the line in front of the Royal as they rolled coins along the pavement towards a designated join in the concrete. The RAAF had earlier played this game, but the better paid Americans used florins, two-shilling pieces. As it transpired, many of the silver coins they now used had been minted in the US Department of Treasury at their San Francisco and Denver mints. From 1942, such was the influx of Allied personnel that Australian mints could not meet the demand and turned to the US.

Riverside Park was a favourite with the Americans, as many strolled in the company of newly found female companions. Visiting VPB-52 aviators were much intrigued by Goat Island, the willow treed landmark in the middle of the Murray River. Almost two years earlier their squadron had been operating from a USN tender based off Goat Island, Jamaica, in the Caribbean. Lieutenant (jg) Edwin "Red" Byrd had been operating in primitive Dutch New Guinea and recalled:

> We arrived at Lake Boga on a beautiful day after passing land that was stark, and straight away landed,

thereafter hastening to make our way for the hotel in downtown Swan Hill. Once on the water, I noticed the water hyacinths [ribbon weed], a nuisance to be reckoned with on landings and take offs. It wasn't a serious problem but rather a frustration of the mind. Once ashore, it took but a short time to enjoy the first bitter ale and plant my feet under the table for steak and eggs, a delightful meal which we had enjoyed while in temporary residence in Perth WA.

We had been flying for about four days and were fortunate to be able to drop off the Catalina seaplane at Lake Boga. The efficient maintenance and repair staff made our chore a simple operation. The following day our crew surveyed Swan Hill streets and restaurants, met [local] government officials while casting curious glances at the lovely ladies to be observed. That was the fun part of our trek through Swan Hill.

The Americans continued their evening cocktail parties in the Commercial Room of the Royal Hotel, delighting their guests with titillating tales of the South Pacific. Ed Byrd:

> I remember of evenings we officers used to give cocktail parties at the hotel and invite your lovely young ladies along. It did not take long to find the Murray River from Victoria to New South Wales. On an afternoon we often walked to the park for a bit of R & R with young ladies who were otherwise not occupied. They were generally athletic, hospitable and always of good spirit. Although the road from downtown Swan Hill went to the tavern, across the river, the tavern [Federal Hotel] was strictly out of bounds to women.

By this date, there were many reports of animosity between some Australian and American forces, primarily over the Americans' popularity with Australian women and their superior spending power; however, this problem was not evident in the Lake Boga-Swan Hill area.

Protected from the morning's chill by their oil skin coats, on 2 August Sergeant Tom Moody and a Marine Section LAC burbled across the misted water out towards USN Catalina #74. Coming alongside, VPB-34 Squadron's Lieutenant (jg) Melvin Essary and crew climbed through the port blister, then readied their aircraft for a 0815 take off. Up and away, a heading was set for Hamilton Reach, Brisbane then on to seaplane tender *Orca* in Dutch New Guinea.

Some months later Lieutenant (jg) Essary was involved in one of the largest air sea rescues ever undertaken by USN flying boats. On the night of 3/4 December the destroyer USS *Cooper* was blown apart by an enemy torpedo during surface action in Ormoc Bay, in the Philippines. Half of the destroyer's crew survived, with most choosing to remain in the water hoping for rescue rather than swim to an area still under Japanese control. Covered by USAAF aircraft, five VPB-34 Catalinas joined in this daylight rescue. Lieutenant (jg) Essary picked up 45 survivors, while another flying boat incredibly took aboard 63 men.

During August, an RAAF Lodger Unit was established at the Boga Depot with personnel arriving on posting for training and eventual relocation to No. 2 Flying Boat Maintenance Unit (No. 2 FBMU), Darwin. These airmen were now involved in the dismantling of the five forlorn Dornier flying boats, lying in open storage.

The 8 August 1944 was a landmark day in that No. 1 FBRD, Lake Boga, was removed from its secret classification. The following day, the sight of CO Scotty Allan resplendent in collar and tie instead of his accustomed roll necked skivvy signalled the arrival of a media contingent which descended on the Depot. For a two full days, the group photographed, filmed and reported on many facets of Depot life. The visiting news gatherers included a Cinesound newsreel team, a party of Australian and overseas war correspondents, plus newspaper and magazine reporters. From 16 August, extensive news coverage appeared in Melbourne newspapers. "HUGE SKY BIRDS AT LAKE BOGA, The Secret Out" *The Age* proclaimed. While graphic magazine features followed, the Cinesound film coverage concluded with Scotty Allan and his Depot crew boarding a VPB-33 Black Cat piloted by Colin Sillers for a take-off from the lake. Three days later Sillers departed for New Guinea, writing of his brief visit:

> We were treated like royalty. I especially remember the cold beer and Happy Hours talking to the lovely young ladies after being away from home so long.

On 16 August, two VPB-52 Black Cats landed two minutes apart. Flying #22 was Lieutenant (jg) Ed Byrd who had been in Lake Boga a few weeks earlier. He was bringing #22 for repairs after it had overflown a USN task force with its

IFF switched off and been damaged by "friendly" fire. Lieutenant (jg) Byrd departed Swan Hill for Sydney by train the following day, recalling:

> To Melbourne via freight-passenger train, took time for us to traverse the distance because of the frequent switching around, back and forth of cattle and sheep as well as other merchandise and produce. It was winter and we were often times chilled from the elements. Stopped overnight at Bendigo at a very nice hotel, drank spectacular wines from Mount Lofty. Left Occidental Hotel in downtown Melbourne, boarded Super Train [*Spirit of Progress*], deluxe carriages. Each weekend men boarded the train on the round trip to Sydney, the sole purpose being to gamble cards and dice as the train was beyond the reach of the law. State owned railroads travelled on different gauged tracks. As a result people, equipment and cargo were transferred at the border between state lines. It was a real mess.

By this date Depot personnel were undertaking the specialised task of converting numbers of amphibious PBY-5A Catalinas back to waterborne only aircraft. The versatility of the PBY-5A had come at a cost, its extra weight. With a total of 46 such RAAF flying boats coming on charge, conversion of a number would enhance their range, performance and load carrying capacity. Selected PBY-5A Cats now had their side and bow wheels removed plus all hydraulic gear. After a series of stout stringers was riveted into position the fuselage wheel wells were flush sealed with duralium sheeting. Likewise, the retractable bow wheel and bay doors were removed, the area reverting to a fixed flush contoured bow. This procedure consumed an average of 1,500-man hours, and reduced aircraft weight by a like number in pounds.

Unusual for Boga in August, at month's end a dust storm enveloped the Depot. This storm was but a harbinger of what would become the greatest period of dust inundation recorded in the area, both in frequency and degree. Over the ensuing nine months the dust menace would severely tax operational efficiency and morale.

From July through November, arrival of USN operational aircraft exceeded those of the RAAF. A pattern had long emerged whereby RAAF aircrew stopovers were generally of shorter duration than those of the Americans. RAAF personnel often returned to their squadron, departed on leave or visited friends or relations. While some Americans journeyed to Melbourne, most stayed in the Swan Hill area awaiting restoration of their aircraft.

Nell Frazer, a telephonist at the Swan Hill exchange, recalled Americans requesting long distance calls home:

> I remember we had young homesick Americans at the public telephones wanting to call the United States. It was not possible. International calls in those days were rare and involved long delays and were usually confined to urgent calls such as sickness, a death or special priority calls related to the war.

At Boga on the morning of 27 September a road convoy of eight heavily laden RAAF transport vehicles departed the Depot, *en route* to the North West Area. Aboard the convoy were personnel of the former Lodger Unit, now

A USN Black Cat undergoes maintenance in the No. 2 Hangar in 1944, while another prepares to depart via the No. 1 slipway in the background. (GU Allan)

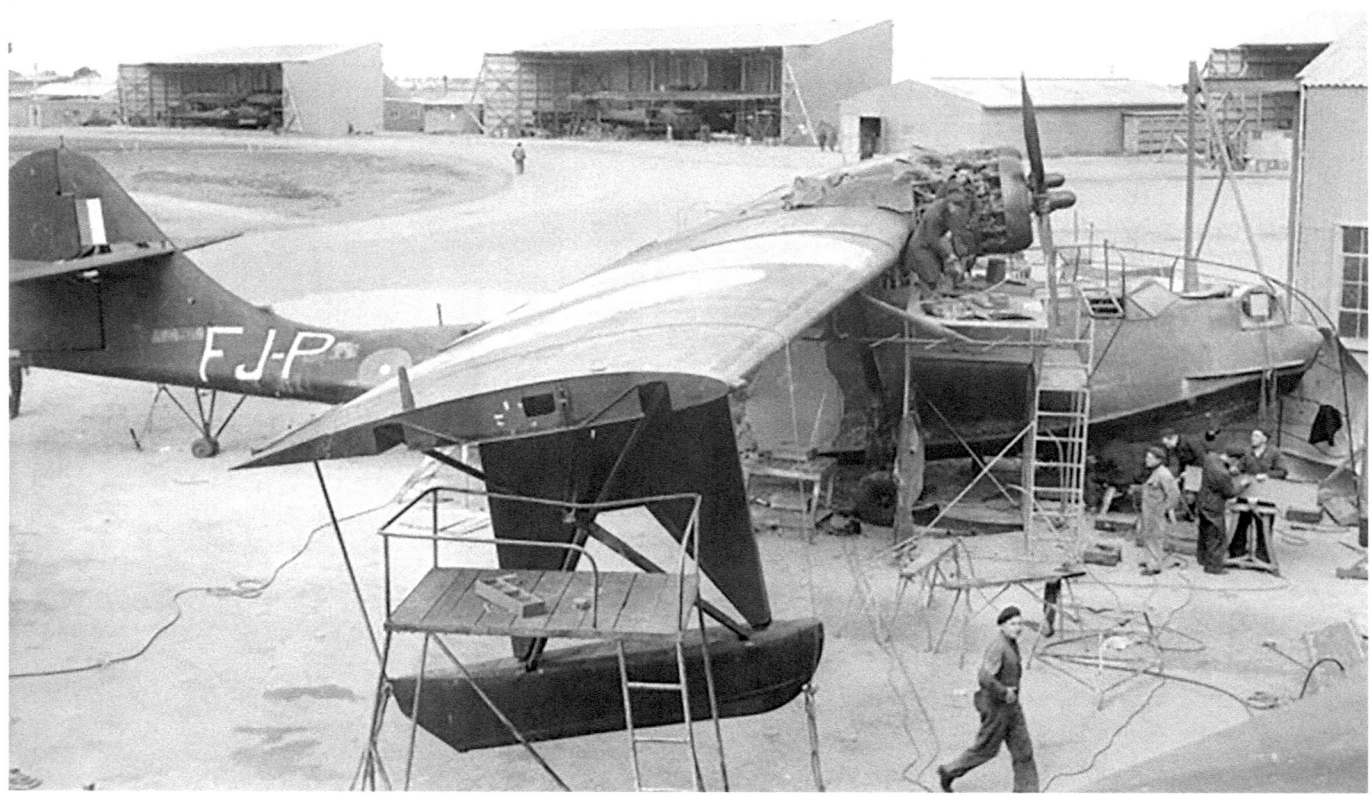

PBY-5A A24-69 of No. 11 Squadron being serviced at Lake Boga in August 1944. The photo gives a good view of the open hangars which were designed so that maximum light was obtained for intricate repairs. (AWM)

Lieutenant (jg) Colin Shillers takes-off for the media from Lake Boga in USN Black Cat 08465 on 10 August 1944. (Ken Chatley)

personnel of No. 2 FBMU, off to Darwin on a gruelling thirteen-day journey through the rigors of central Australia. Late that afternoon a further dust storm hit the area followed by rain providing aircraft, vehicles and buildings with a mud-dappled dusted finish. Try as staff may to remove it, dust bedevilled the Depot. A signal was received from Cairns expressing concern and enquiring what exactly was going on down south, as newly received aircraft arrived in less than pristine condition.

However, during the life of the Depot amenities had slowly improved. Up at the camp September saw South Australian ACW Kathleen Doody open a WAAAFs hairdressing salon next to the airmen's hairdressers. Kathleen soon perfected the Boga Bob, the Cat Curl, the Mariner Mop and the Sunderland Sweepback. Plants had struggled up the fronts of WAAAF living quarters, while newly grassed areas around the camp showed some promise.

Additional to work already on hand at Boga, October saw No. 1 FBRD undertake the modification of B-24 Liberator bomber and Mosquito fighter bomber parts, work that was executed in "fast time". By now a large gate had been constructed to the north-western perimeter fence, through which flying boats could be towed to an adjacent paddock which had already been levelled in preparation for the marking out of a compass and DF loop swinging area.

In order that these vital aircraft navigational and radar instruments might be accurately adjusted away from the interference of metal objects, Department of Interior surveyors had arrived, setting out a semi-circular range with a radius 800 feet. Markers were then driven around the circumference at five-degree intervals, commencing at 090 degrees, and terminating at 270 degrees magnetic. Calibrations to sensitive equipment aboard aircraft would soon be accurately adjusted from an oscillator fixed to a concrete platform now under construction at the centre of the large semicircle.

Dutch flyers Lieutenants Henk Hasselo and Rynders and crew had arrived from Sydney via Albury and Murray Valley Coach to pick up Catalina Y-87. This Cat had arrived in three weeks earlier from Lake Santani near MacArthur's Hollandia Headquarters. Henk Hasselo recalled this visit:

> We were about to take Y-87 on an acceptance flight when some of the Boga mechanics said "How about giving us a buzz". By way of saying thank you for the job they had done on our aircraft, at the conclusion of our test flight we shot up the base. We made a couple of very low runs and steep turns when we were ordered down by radio. Scotty Allan matted us indicating had he jurisdiction over us he would have had our wings.

October 3 and 4 saw three USN Catalinas depart Boga for the New Guinea area via Brisbane. As the Americans were staging through the Queensland capital, No. 41 Squadron's Noel McKnight, now a flight lieutenant, had just departed the Brisbane River in Mariner A70-2 bound for Lake Santani. McKnight had on board 44 US WACs (Women's Army Corps). Before take-off, with their gear, these forage capped, neatly uniformed shirt and slacked gals, had settled themselves into the long, drop-down canvas seats, affixed to the Mariner's fuselage. Flight Lieutenant McKnight:

> Having manoeuvred the aircraft along the confined waters of the Brisbane River, at Hamilton Reach, I gunned A70-2 and lifted off under full power, heading west northwest on an eight-hour journey to Hollandia. On this long flight, our passengers eventually found the need to visit the bathroom. Suddenly screams and peals of laughter erupted from the central unscreened section of the aircraft as many of the WACs struggled with the small funnel and hose device normally reserved for male ablutions.

No doubt influenced by the sight of US Navy Catalinas fitted with quad 0.50-inch calibre machine guns, the Dutch now arrived at Boga in Y-45 requesting the installation to their aircraft of similar high-powered armament. A month later, fully serviced with its four new machine guns securely fitted within its bow cavity, Lieutenant Leeflung departed Boga, mentally rehearsing the saddle mounted control switching that would give him awesome fire power should the occasion arise. As it transpired, Dutch aircraft visiting Lake Boga did not make contact with the Japanese.

During October, the WAAAF held their first "At Home", returning hospitality that had been so freely extended by local residents. However, the month's social activities were quickly curbed. Palma Kenzor (McCallum) recalled:

> On Saturday, girls in our hut rode bicycles down to Mystic Park for a picnic. Late in the afternoon we returned to our huts full of fun, but were immediately hushed, then informed that WAAAF officer Kelso had drowned

Armourers hoist a mine into position on an RAAF Black Cat from a bomb scow. A portable winch is hand wound from atop the mainplane. (AWM)

in the lake that afternoon. They were still looking for her, I went to the door and could see a Catalina circling, searching.

Section Officer Norma Helen June MacDonald Kelso from Tarrenlea, near Hamilton, Victoria, had arrived at Lake Boga on posting to Cypher duties just eleven days earlier. This tragedy had occurred while the officer swam from a boat accompanied by Depot personnel. The Sunday morning ushered in a mantle of gloom as Section Officer Kelso's body was recovered at 1000 hours in 14 feet of water. A funeral was held the following Tuesday, with Kelso buried at Lake Boga War Cemetery.

A fire broke out in the inflammable store mid-month, but quick action by the fire picquet saved the situation. Three days later the weather again worsened with more dust, the Unit Operations Record Book, noting:

> October 16. Adverse weather conditions, with high winds, heavy dust storms continued throughout the day. Members reporting on posting had to be conveyed by buses from Kerang, owing to a goods train being derailed by sand on the railway line. October 17. The dust storm continues. This storm is worse than the one on 16 October.

Further north, October saw the Battle of Leyte and the American return to the Philippines. This ultimately paved the way for RAAF Catalinas to be used for ultra-long range mining missions in the last year of the war, as far as the Chinese coast. Interestingly, one ex-Lake Boga airman participated in the Leyte invasion. In 1942, LAC Clifford Moulton had been among Boga's early RAAF arrivals, and was mustered to administration. He was first billeted at the Royal Hotel, then with his wife and son "lived out" with the Greenham family before relocating to the Depot camp. In August 1943 Moulton departed Boga on a posting to Brisbane where he undertook a specialised intelligence course and was subsequently among the 24 combined RAAF, AIF and British Army personnel who formed the Wireless Intelligence Kana Interception Group. They were attached to the US Fifth Air Force during MacArthur's Philippines invasion, and quickly established a top-secret intelligence station to listen to Japanese radio transmissions.

In the early morning of 4 November, still operating three shifts over 24 hours, the signals section received a message requesting permission to alight. A Catalina was diverting from its planned Perth to Rathmines course. The damaged craft had flown through the night and was low on fuel as No. 20 Squadron's CO, Squadron Leader Athol Wearne switched to a south easterly bearing and by 0800 hours had Lake Boga in sight.

Ground staff watched and wondered as Catalina A24-60 descended. Something was amiss, and the craft had only one of its floats down. Concern mounted as it became apparent that the port float was missing and the port mainplane seemed marginally foreshortened.

Down down, to a steady alighting, then Squadron Leader Athol Wearne's waterborne skills came into play as he taxied toward a mooring buoy. Motors cut but still under way two crew members suddenly climbed from a front hatch and were up on the mainplane, crawling to the extremity of the starboard wing, their action weighing down the existing float and so avoiding the possibility of the Cat "turning turtle". The situation under control, Wearne came ashore and advised CO Allan that he and his crew would continue their flight to Rathmines having once refuelled and partaken of break fast. 'Aye, but no,' exclaimed Scotty 'Ye cunnot proceeed, the aircraft is unsuvicable, it ornly huss one float.'

Two airmen from No. 1 FBRD take out one of the unit's yachts on Lake Boga, a popular summertime activity. (AWM)

The exasperated Wearne bit his lip as he thought to himself "Of course it only has one bloody float. I should know. Didn't one of my pilots clout the mast of a trawler as he came into Exmouth Gulf, losing a float and damaging the mainplane. Haven't I flown from Darwin to Exmouth, inspected the aircraft, flown the machine to Perth, then by night flown across Australia, limping along at 80 knots only to reach Lake Boga and be told by a determined Scott. A YE, BUT IT ORNLY HUSS ONE FLORT!"

Acknowledging the right of the CO of a RAAF engineering Unit to terminate the aircraft's journey until restored, Squadron Leader Wearne, with a controlled limp, walked off to make enquiries regarding an alternate flight to Rathmines. A controlled limp? Correct, for Wearne was a somewhat unusual CO and Catalina Captain. He had only one leg, having lost the other to a shark attack while serving at Cairns in September 1942. After twelve months rehabilitation Wearne returned to active service flying throughout the war; and beyond, with the use of an aluminium appendage to which was attached a most useful bottle opener.

Warmer weather saw the Depot's eight yachts, once again refurbished, sail the lake in leisure hours while some personnel resumed district tennis and cricket. A popular picnic spot in the Swan Hill area was Goat Island. Named for the small herd of goats that earlier frequented this landmark, the prevailing drought conditions and consequent low river levels had made Goat Island readily accessible by foot from the New South Wales side of the Murray River.

For some time, tales had circulated regarding wild parties held on the island, attended by both local residents and service personnel. Late in 1944, a group of young boys, myself included, went bird nesting on the island and in the process came across a pair of quality khaki under shorts branded USAAF with patent and product numbers clearly stamped in black. How on earth would an Allied aviator forget such an important item of apparel, we young lads pondered?

The answer to this vexing question became a little clearer the following year when I attended Melbourne Grammar School in St Kilda Road. The US Army had recently vacated our boarding house. Adjacent to the school stood Kellow House which had, pre-war, proudly distributed Woolsey, Packard, Bentley, and Rolls Royce motor vehicles. However, the intervening war had created an empty show room and the building had long been commandeered by the RAAF. Kellow House had in fact, until mid-1944, been the home of No.4 (Maintenance) Group, the engineering arm controlling No. 1 FBRD, and it continued to house RAAF administrative personnel.

On some evenings, from the top floor of the school's Perry House dormitories, we students observed RAAF and WAAAF personnel, late in leaving Kellow House, locked in deep embrace against the trunks of St Kilda Road's great elm trees. On occasions our parents' hard-earned pennies or half pennies were dropped down through the branches to the pavement below. Upon hearing a clink of a coin strike the pavement, the embracing couples broke free, patting their pockets then searching the darkened footpath for the imagined dropped coins. "Anyhow" mused we sniggering students "why did these airmen take so darned long to kiss these WAAAFs?"

In November three unlikely vehicles arrived at Lake Boga in the form of DUKWs, large army amphibious landing craft. On November 6 two vehicles arrived under the direction of Lieutenant Graham while RAAF Marine Officer Flight Lieutenant De Castella and party drove in four days later with a third vehicle. Commonly called "Ducks", the significance of DUKWs was as follows: "D" a General Motors classification for the year of manufacture (in this instance 1942), "U" for Utility, "K" meaning front wheel drive and "W" meaning two rear driving axles.

Among other tasks the DUKWs were tested as an alternative to launches as supply craft to flying boats but proved too unwieldy for operation in the vicinity of the thin-skinned aircraft. Regardless of practical considerations, the DUKWs proved most popular during recreational hours, as recalled by the then Kangaroo Lake resident; now Air Vice Marshal (Ret' d) Alan Heggen AO. In a contribution to Ogilvy's *Riverboat Newsletters* AVM Heggen wrote:

> The DUWKs were known to appear in some unusual places in what might be termed "aid to the civil community tasks", including the retrieval of upturned pleasure craft, provision of logistically well-equipped fishing platforms and I have no doubt, various other roles of which a 12-year-old schoolboy would not have been aware.

At the 11th hour on Saturday the 11th of November, camp and Depot sirens signalled two minutes silence in observance of Armistice Day. That afternoon, personnel listened in to the running of the Melbourne Cup. Before mess on Monday evening the Depot CO was glancing through the pages of the daily newspaper when his attention was drawn to a photograph of one of the Depot's personnel at the Melbourne Cup, an attractive lady on either arm. On entering his office the following day, a sixth sense caused the CO to enquire into any leave granted to photographer Gordon De Lisle. Sure enough De Lisle had journeyed to Melbourne for the cup with only a local pass. De Lisle was confined to barracks for six weeks and, for a period, was deprived access to the Photographic Section.

Also in November the first of the RAAF's new waterborne only PB2B-1 and PB2B-2 model Catalinas had arrived at Boga. Built by Boeing Aircraft in Canada, the PB2B-2 Cats were readily recognisable by their higher tail fin. This model was also fitted with improved radar and more powerful Pratt and Whitney engines. The Canadian aircraft had been received by the RAAF at San Pedro, California.

A change of command. On December 8, three days before assuming command of No. 1 FBRD, Wing Commander Raymond S Rice alighted at Lake Boga aboard Catalina A24-27. Within a month, Wing Commander Rice would receive a posting to the RAAF Staff School, nevertheless it seemed appropriate that the new CO's arrival should be heralded by dust storm, a storm similar to that xperienced days previously which "had obliterated the area with lights having to be turned on".

A pilot, with both an engineering and science degree, Wing Commander Rice had joined the RAAF in September 1939, No. 1 FBRD being his first command. The outgoing CO Allan was prompt to conduct his successor on a tour of inspection. Allan departed No. 1 FBRD on 11 December on posting to No.2 Personnel Depot (Discharge Section).

At 1320 hours, down at the jetty, Cat captain Squadron Leader Bryan Monkton and crew were already aboard a Depot launch as its engine burbled away in neutral. Having farewelled the assembled officers, Scotty Allan eased himself into the waiting craft, looked up, and with a wry smile saluted, as the launch rocked away towards new PB2B-1 model Catalina A24-202. Squadron Leader Monkton:

> On this flight I was officially the captain although, as I understood this was the last flight Scotty Allan would make as a member of the RAAF, I invited him to do the flying, an invitation he unhesitatingly accepted.

On December 14, No. 1 FBRD CO's supplementary diary noted:

> A PBY-5A was received from the US Army, for repair but was not serviceable for a water alighting. It therefore landed at the Swan Hill aerodrome where it was partly dismantled and then towed to the depot by road.

Lieutenant Wentjes of the US Army's Fifth Air Force had landed at the Swan Hill 'drome on 6 December with severely damaged Catalina #433885. The hull was shattered and bore testimony to the severity of a waterborne alighting in which this aircraft had recently been involved.

Depot personnel soon removed the two outer mainplane panels and transported them by tender to Lake Boga. This manoeuvre was followed by the unusual sight of a tail towed, clipped winged, Catalina amphibian progressing slowly along the Murray Valley Highway, bound for an eight-week restoration.

By way of diversion, on 19 December, two Kittyhawks on exercise from No. 2 OTU at Mildura landed on the salt flats near the Depot barracks while a Wirraway landed by the Transmitting Station. The P-40s aircraft were towed to the Swan Hill aerodrome where they were "refuelled and allowed to proceed" while the Wirraway was flown from the stubble paddock.

The dust menace was really taking its toll as a litany of dust related problems appeared in official Depot reports. Over the three previous months, one thousand one hundred sick parade visits had been recorded by Depot MOs, all due to dust related eye problems. Although goggles were issued to those operating out of doors work was still difficult. In spite of support from No. 4 (Maintainance) Group, appeals to authorities for funds to line and semi dust proof buildings, were continually turned down.

The armament section had to rework machine guns resulting in a twenty percent increase in the time factor. As Christmas approached, dust seeped into open fuel lines, starter motors, generators and all things mechanical, doing nothing at all for morale. A measure of the moving Mallee soils can be judged from a report of 18 December:

> No. 4 hangar. 1,400 pounds weight of dust was swept from the wings and fuselage of a PBY and the open hangar floor area. A24-56, 210 pounds swept from aircraft. Had to be totally recleaned.

The dust had also proved a dampener to personal relationships between RAAF and WAAAF as an item in the Depot's news sheet *Bogafobian* indicated:

> SYMPTOM OF BOGAFOBIA ...
>
> Pleasant day - arrangements - service type blanket - stones on roof - whistling - lake destination - grass and insects - aroma of reeds - arrival of dust storm - THE END.

December had seen a further advance in Depot communications with the erection of a 50-foot radar tower adjacent to the Depot entrance, while on the recreational scene a new boxing ring was constructed to professional standard. On Christmas Day 1944 at the Lake Boga Airmen's Mess, it was the Depot's third Christmas Dinner with officers and sergeants again waiting on other ranks.

During 1944 more than 170 flying boats had alighted at the Depot most of which had been serviced or restored. This year of change in which the Allies had gained the upper hand in both the European and Pacific theatres of war, closed with No. 1 FBRD at peak strength of 943 personnel. They comprised 39 officers, 802 airmen and 102 WAAAFs.

One of the huge dust storms over No. 1 FBRD's accommodation huts.

CHAPTER 4
1945

With the European war in its sixth year and the war against Japan in its fourth, No. 1 FBRD personnel continued their duties, all contributing within their varied musterings, towards the restoration and return of Allied flying boats to forward areas of conflict in the western Pacific.

A landmark occurred on 2 January when the last USN Catalina departed Lake Boga This was Black cat #27 from VPB-33 and was piloted by Lieutenant Commander James Merrit Jr. It departed for the Philippines.

Early January also marked the end of a long-term rebuilding job. RAAF Catalina A24-65 had been so badly damaged that it had arrived at Boga by rail from Darwin in several crates in March 1944. After almost a year it had been fully restored and test flown, prior to its departure on 8 January for receival by No. 43 Squadron.

Squadron Leader Rice's reign at No. 1 FBRD had been brief, with Wing Commander Roy H Foord assuming command on 7 January. This new CO had been appointed to a RAAF commission in April 1940 and now toured the Depot area, familiarising himself with aspects of its operation. Within two days, Foord was greeted by a "very bad" dust storm, and he then received a signal in reference to an escaped German POW when all necessary precautions were taken.

Still based at Darwin, No. 20 Squadron's CO, Squadron Leader Athol Wearne, was again involved in the retrieval of a disabled flying boat, on this occasion Catalina A24-87, an aircraft that had departed Lake Boga some six months earlier. On 9 January it had been forced to return to West Bay near Learmonth when the starboard engine lost power after a mining operation. Wearne later recalled:

> There being no facilities for an engine change at West Bay and little chance of recovering our aircraft for some considerable time, I suggested to the senior Engineering Officer at North West Area that I attempt to fly her out on the port engine, and this was agreed.
>
> The recovery of this aircraft required a reduction in weight with the removal of all extraneous equipment. A decision was also taken to remove the starboard propeller in order to reduce drag, while the amount of fuel carried was lowered to 300 gallons, sufficient only for the flight back to Darwin.

With Warrant Officer Brown as second dick, the crew balanced the aircraft as Squadron Leader Wearne flicked his yoke-mounted switches and called his flight engineer to initiate engine start. The port engine sprang into life. Now, a testing time for the PBY and its only firing Pratt and Whitney as each crew member asked himself the same question. Can a take-off be achieved by a one engined Catalina? Squadron Leader Wearne:

> Unfortunately there was very little wind and the bay was very calm. These features, together with the heat, made things more difficult than would be the case if a decent breeze was blowing and a good chop on the surface was available. As a result the take-off run took four minutes. The aircraft seemed to just not want to stay up on the step, no matter how hard I worked. Eventually enough speed was gained and we came unstuck much to the relief of all concerned especially the flight engineer who was very worried about the excessive temperature of the engine. Once airborne the aircraft performed admirably and we arrived without incident, back at East Arm [Darwin], after a four-hour flight. Although overheated for more than the maximum time advised, no apparent damage was done to the engine.

Corporal Jack Settle had seen four summers at No. 1 FBRD, and in April would marry local schoolteacher Gwen Williams. During those austere days when personal possessions were few and most highly valued, he recalled an incident which occurred while working on a waterborne Catalina:

> During the summer of 1944-45, severe bushfires in southern Victoria destroyed an airman's home. Although

his family survived, he lost everything. We were out on the water, he was standing with us on the portable catwalk working on a Catalina motor, when his watch band broke. This poor fellow was not going to lose one of his last valuable possessions, so he dived in after his watch. In his haste he overlooked the fact that he had several spanners and other heavy tools tied to his wrists and belt (a common practice so he would not drop them in the lake). We dived in and hauled him out of the water, he had no chance on his own. We saved the airman but unfortunately, not the watch.

The "Dust Monster" of 30 January 1945 about to hit a hangar with a resultant blackout over the Depot.

Dust again devastated the district on the 22nd and 24th of January. On Tuesday 30 January the daddy of them all arrived. Amidst searing heat, a fierce dust storm blew day long howling through the camp, Depot and district. Then at 1730 hours it further deepened in intensity to a total blackout, obliterating the area until 1900 hours. The *Guardian* later reported:

> THE DUST MONSTER! Worst Dust Storm On Record! Day turned into night, as the dust monster went on its wild rampage burying everything beneath its mighty mantle. Tuesday 30 January was indeed a shocking day.

Each with his own headphones, Squadron Leader Courtney Oldmeadow and crew members board Catalina A24-306 at Lake Boga on 21 February 1945.

In recent months the RAAF had lost several Catalinas in the northern area, including A24-204 which had departed Lake Boga on 5 January and was lost over the Timor Sea three weeks later. As a result the USN offered the RAAF a number of Catalinas which were buoyed up at their Manus Island base. Sergeant Richard Clapson was one of a group of personnel selected to secure the aircraft:

> I and 21 others flew from Lake Boga to the American base at Manus Island to collect a number of Catalina aircraft (I think six) which had suffered some damage and had been offered to the Air Board by the Americans, free for the taking. We all flew up in one Catalina but when we got there the lot had been sunk as they thought we were not interested.

During January eleven flying boats had arrived at the Depot while thirteen had departed. On 1 February, Squadron Leader Jock Tennant assumed temporary command of the unit as CO Wing Commander Foord departed on leave. That evening punches flew thick and fast in the new boxing ring as another great fight night was staged at the Depot Recreation Hall with the main bout sponsored by Ma's Boomerang Cafe. A record attendance of 700 cheered the fighters, the event being so well patronised that many had to watch proceedings through windows.

With its distinctive high tail fin, a new Boeing PB2B-2 Catalina that had arrived in from Canada on 17 January was now operationally fitted: the pilot's armour plating was removed, under wing mine racks were in place and electrical wiring complete. The aircraft now rolled down the slip on its beaching gear preparatory to test and departure. Part of a diverted RAF order, this aircraft had arrived as JX623 but now bore RAAF nomenclature A24-306. Skippered by Squadron Leader Courtney Oldmeadow, A24-306 departed Lake Boga on 5 February to be received in Darwin by No. 20 Squadron. Weeks later this flying boat would be among RAAF Catalinas mining Hong Kong Harbour!

The dust problem continued into the heat of February. On 12 February it was noted "Very bad dust storm all day, visibility 100 feet". While predictions of at least another full year of war were made, the Minister for Repatriation CW Frost visited the Swan Hill district to examine plans for post war soldier settlement.

RAAF Catalina A24-206 taxis across Lake Boga in February 1945 prior to delivery to No. 43 Squadron. This flying boat had recently arrived from the US before its conversion at Lake Boga to a Black Cat. Note the raised radome above the cockpit. (AWM)

By this date the RAAF Nos. 112 and 113 Air Sea Rescue Flights equipped with Catalinas were operational. The latter had relocated from Cairns to Morotai where, until March 1945, it was attached to the 2nd Emergency Squadron of the US Thirteenth Air Force. Known as the *Snafu Snatchers* this US squadron had rescued approximately 300 airmen at the time of the RAAF arrival.

The badly damaged USAAF Catalina that had arrived in at Swan Hill in December was now restored and fully operational. Accordingly, 13 February saw the penultimate departure of a US military aircraft when Lieutenant Wentjes and crew sped across the lake and were away aboard USAAF Catalina 433885. A week later, on 20 February, the last US flying boat departed Lake Boga. This was another USAAF amphibian, 433876, which flew to Rathmines via Sydney. Catalinas bearing the distinctive "bars and stars" would never again be seen at Lake Boga.

Nineteen months had passed since the Yanks had first landed in the area and together with visiting RAAF and Dutch air crew, had added a dimension of reality to a war being fought beyond Australia's seaboard. US aviators had certainly contributed to the fabric of local wartime activity. The Americans were gone but would certainly remain well remembered by Depot personnel and a multitude of district residents, not the least of whom were many young ladies.

New Catalinas kept coming, one amphibious Cat and eight new Canadian Boeings were flown in during February.

Having returned on leave, CO Wing Commander Foord now made a further request to No. 4 HQ (Maintenance) Group for provision of lining materials to messes in an endeavour to ameliorate the dust problem:

> The dust storms have been such that eating conditions in the sergeants' and airmen's messes have been almost impossible. The dust storms in this locality are increasing in frequency and last for twelve hours or more.

Recollections of several airmen confirm that it was standard practice during storms to sit one's saucer on top of the cup and so avoid drinking gritty tea. In a June 1945 communication to the Air Board, No. 4 Group's Air Commodore AW Murphy was to further support the installation of lining to buildings, advising:

> Definite evidence of food contamination could be provided about every third day during the months of November to March inclusive, in the form of 600 odd platefuls of "Meat and Vegs" liberally covered with Lake Boga dust.

Countering a suggestion that lined buildings would more readily house vermin, the Air Commodore replied:

> It is considered living conditions in this locality would be too hard for them anyway.

Meanwhile Catalina A24-109 had departed Lake Boga 15 March, to return to No. 113 ASR Flight. On 29 March Flight Lieutenant Wally Mills successfully retrieved a downed Kittyhawk pilot from waters off Dutch New Guinea. Two days later, now in A24-109, Mills flew a rescue mission after Beaufighter A8-33 was hit while strafing an airfield south of Liang, before ditching in the Haroeka Strait. A24-109 had just alighted close to the downed crew, when enemy shore batteries and camouflaged watercraft opened fire. Within three minutes the Beaufighter crew was on board, and Mills was up and away. One man was severely wounded in the abdomen, while the Catalina suffered only slight damage.

During April, the Depot undertook the manufacture of rudders for P-40 Kittyhawk aircraft, while the parachute refolding procedure received special attention. Yet another clean-up was in progress after a 24-hour dust storm.

A Catalina is towed up the Lake Boga slipway by a tractor.

Within days of departing Boga in March, Flight Lieutenant Bullman found himself involved in one of the most demanding RAAF air-sea operations of the war. Early in April, the Japanese light cruiser *Izuzu* had been shadowed to Koepang. When the cruiser left port a bombing attack was mounted by RAAF B-25 Mitchells and B-24 Liberators. During the attack, Japanese fighters downed two of the Liberators, with eleven crew seen to parachute from their doomed aircraft. Bullman, responding to the rescue call in A24-54, alighted in the area and was retrieving four airmen when attacked head on by an enemy aircraft. His Catalina immediately burst into flames while the crew dived into sea. The aircraft sank within two minutes.

A covering ASR Liberator then dropped two five-man dinghies, into which nine of the survivors clambered. Flight Lieutenant Corrie's Cat A24-58 now alighted in the open sea in this desperate ASR operation. Survivors were quickly loaded through the aircraft's bow hatch. However, two airmen still remained out of sight, their position indicated by the covering Liberator's low passes. By now, Corrie had the rescued Bullman with him up front flying as second dick. As A24-58 taxied in the direction of the missing men two twin engine Nakajima Irvings attacked the Catalina with phosphorous bombs. Corrie hit the horn button signalling "cut the drogue", as the starboard blister gunner pounded away at the attacking enemy. With both blisters wide open the salt spray flew as a full throttle take-off was achieved. Hugging the sea, the laden Cat twisted and turned for fifteen minutes as the Japanese continued their relentless passes. With motors roaring, Corrie then found 3,500-foot cloud cover and safety.

Throughout the operation a Liberator A72-72 had successfully covered the Cat's tail. A search for the two missing airmen was flown until dusk. Sadly, they were never found.

Over two years had passed since a four-engine flying boat had alighted on Lake Boga. April 9 heralded the arrival of RAAF Sunderland A26-1. On an Australia wide War Loan tour, the Sunderland was captained by former Nos. 10 and 461 Squadron's Flight Lieutenant Max Mainprize DFC who had gained his decoration after bagging a German U-boat in the Bay of Biscay. The following day, Marine Section personnel were busy ferrying Third Victory Loan investors out to the giant aircraft, prior to it thundering across the lake and flying away on promotional excursions.

Several times that day the Sunderland thundered across the lake. By 0630 hours the following morning, bird life had barely stirred when Mainprize fired up again. Taxiing to its take off point, the great rudder swung, port engines throttled as the aircraft wheeled round and steadied, then with a power increase, roared across the water rising on its step and lifting on its journey to Western Australia. This sight brought back vivid memories to several Depot personnel, memories of their service in a greener Great Britain.

Within a week of its departure the Sunderland returned from Perth via Adelaide and again engaged in war loan flights. By its departure date, 20 April, 198 passengers had flown in the aircraft having invested more than £120,000. Depot personnel supported the Victory Loan to the extent of £11,840.

Towards the end of April, a flight of four new Canadian built Boeing PB2B-2 Catalinas began their descent upon the lake. More of the aircraft originally destined for the RAF but diverted to Australia, these Catalinas still bore RAF serial numbers. The new models were readily recognised by their distinctive high tail fins.

Flight Lieutenant Lance Steadman was last to alight in JX841, an aircraft still displaying the name David Hornell VC, clearly painted on the cockpit's sides. The last of 362 Catalinas built by Boeing Aircraft Corporation of Canada, this aircraft had been named for the first Royal Canadian Air Force airman to receive the Victoria Cross. Within days the Depot paint section had switched the aircraft's misty blue-grey camouflage to matt black, an RAAF signwriter had applied the number A24-360 and operational fitment was complete. The aircraft was then ferried out for receival by No. 43 Squadron.

Visiting Walrus W3085 taxis through ribbon reed at Lake Boga in June 1945, with a No. 11 Squadron Catalina in the background.

Frustratingly, the first day of May saw more dust. Better news followed a week later on 8 May when banners at Ollie Mills' newsagency proclaimed the news "WAR ENDS IN EUROPE". The following day, Depot personnel were stood down in celebration of VE (Victory in Europe) Day. In Swan Hill, churches combined in a 9am service of thanksgiving at the Town Hall. While a deep feeling of relief prevailed, Japan remained unconquered.

New Catalina A24-306 had departed Lake Boga for receival by No. 20 Squadron in February. By early April, together with other RAAF Cat crews, Squadron Leader Miedecke was quartered at the Jinamoc US naval base, readied for further "Popsy" coded mining operations. Flying the new Cat, Miedecke had successfully mined Hainan Strait (14-hour mission), Amoy Bay (15 hours 15 minutes) and the most demanding of all, Hong Kong Harbour in a 16-hour operation. Switching to A24-83 in May, Miedike and crew mined Hong Kong on two further occasions.

A member of Miedecke's crew was 20-year-old Sydney air gunner LAC Jim Jenvey, who would later settle in Swan Hill. To this day Jim retains a high regard for both his skipper's skills and the qualities of the Catalina flying boat:

> I was always keen on guns and shooting and became an RAAF air gunner. Apart from the .303s and .50 machine guns, we always carried two tommy guns with us while the captain wore a .38 sidearm. We mined many harbours, Hong Kong harbour three times, always with a terrific navigator. Well out from Hong Kong, which was all blacked out in those days, our Cat would drop down to a few hundred feet. As we came in, sure enough the promontories would show up dark against the water, spot on, just where we expected them. The amount of shipping we could see in that harbour was amazing. By this time, our navigator was up in the bow ready to release our mines at a given height and a given speed. Each time we flew right down the planned line, from datum to datum, then out and home.

During May, Dutch Cat Y-45 had flown in with a crew for Y-87. One Mariner and fourteen Catalinas had arrived, seven of which were new Boeing models. Twelve flying boats had departed.

The weather at Lake Boga on 23 June was deteriorating, and it would soon gust to 25-30 knots. Accordingly, Wing Commander John McMahon made an early departure for Schoefield aboard the visiting amphibious Walrus W3058. Starting the old "scow" was quite a task. With the mooring line slipped and drifting, a crewman was seen to hoist himself from the cockpit, up on to the lower wing and with a crank handle, briskly wind the engine's fly wheel until the accompanying whine reached the required high pitch. At that point, McMahon pulled a toggle, engaging the impulse starter. Mixture set, the 775 hp reverse facing Bristol Pegasus engine spluttered into life as the crewman ducked back into the fuselage. Sailing down wind, McMahon finally swung his heavy control wheel to bring the craft about, then with his left hand, pushed on power. The Walrus rocked across the waves, the captain ensuring his aircraft was stuck down tight until 70 knots was achieved. Then back with the control wheel and the Walrus was away.

June closed with the Operations Record Book noting "Gale weather, rain and squall all night, Engineering Officer, Deputy Engineering Officer, and Duty Officer stood by aircraft during night." The Depot strength in mid-1945 stood at 897 personnel.

On 18 July a signal was received advising Mariner A70-2, with a badly damaged bow, was on its way from Cairns. What an understatement! At 1500 hours Flying Officer K Roberts alighted in the faded flying boat and proceeded to buoy up. To Marine Section personnel already on the water, the damage became apparent. Much of the Mariner's bow was covered in crudely riveted sheets of galvanised iron. As one launch brought the crew ashore, the other was out releasing the Mariner and tail towing the craft to slip No. 1. In wading gear, personnel then moved into the frigid lake and attached the cumbersome beaching gear.

Flight Lieutenant George Mason's Mariner A70-2 after its collision with a large beacon at Cairns.

As a tractor slowly slipped the enormous bird, word spread as airmen gathered to inspect the aircraft. There for all to see, was the Mariner's bow sheathed with Lysaght sheeting, proudly displaying its Queen's Head brand and noting the sheet metal's quality: "War Finish". A month earlier, during a night take off at Cairns, this Mariner, captained by Flight Lieutenant George Mason, had experienced a near disaster. Mason described how the aircraft had struck a huge unlit channel marker:

> There were a series of multi piled channel markers, as big as houses, out from Cairns. Having passed the last lit marker, preparatory to take-off, I had just given my Mariner the gun when the flying boat crashed into the final unlit beacon. With an unholy crunch our bow collided square on and concertinaed almost to the cockpit bulkhead. The bow had acted as a perfect shock absorber. Remarkably we were able to taxi back to our mooring. Subsequently sheets of galvanised iron were affixed to the bow and the aircraft was successfully flown to Lake Boga for repair.

The bow of Mariner A70-2 sheeted in "War Finish" galvanised iron, as flown from Cairns to Lake Boga.

As A70-2 was being towed to a holding area, a light tender departed the Depot for the Swan Hill aerodrome where four Tiger Moth aircraft had landed *en route* to Mallala, SA. Guards were mounted on the aircraft and the pilots were driven to the Boga camp where they were accommodated overnight.

At 1010 hours the following day Flight Lieutenant Lee McKeand alighted in A24-371. McKeand recalled his approach to Catalina operations:

> One was given his operations order and then with the appropriate crew members calculated various requirements ... a multiplicity of tasks, fuel, bomb weight and numbers, mix of ammunition depending on expected activity in an area, tracers, armour piercing shells in machine gun belts ... correct maps, weather report, wireless contacts, all this mostly carried out by fellows twenty-one- to twenty-five-years-old.
>
> We balanced our aircraft before take-off, the Flight Engineer had a small spirit level fixed to his tower. In the tropics it was important to lift off on the first run as engines at full throttle overheat and take a long time

to cool for a further try. Once away no thought was given to danger, you were busy reading your aircraft in every quarter in preparation for the task ahead. You did all you could to successfully reach the target, relying on your previous experience. Over the target you worked to avoid known hazards. The speed, clarity and flexibility of one's mind when flying through a target was nothing short of astonishing. Mission accomplished, your aircraft up and heading for home, a new dimension took over, elation, exhilaration then a further wave of adrenalin powered emotion, swept through each crew member.

Then fatigue took over, one often faced an eight-hour flight back to base, sometimes the captain would hand over to his number two, then with aching animal tiredness, grope his way aft to a waiting bunk, and deep sleep, then be shaken by a crew member: "Skipper, there's a strange light down to starboard, come up and see what you think."

Just days to the war's end and another Boga airman was killed. On Sunday 29 July Flight Sergeant Allan David Tozer, a passenger in a Great Gatsby styled single seater Packard, died instantly when thrown from the roadster after a collision with another vehicle.

Nearing the end of a long flight, Ceylon to Sydney, Dutch Catalina Y-69 carrying two crews paused briefly at Lake Boga on 1 August. The Captain, Lieutenant Claterboss had flown via the Cocos Islands, Perth and Adelaide. Upon arrival in Sydney, several members were posted to Tocumwal to commence conversion to B-24 Liberators, an aircraft which the Dutch were operating out of the Cocos Islands on bombing missions, convoy and submarine surveillance.

On 3 August, the last new Catalina to arrive in Australia in wartime touched down on the lake as Flight Lieutenant Armand Etienne arrived in A24-384. War's end was days away. It was a crisp clear moonlight night at Lake Boga, at 0245 hours on 6 August. Rugged up against the cold in his great coat, a drowsy RAAF guard stood silently in the sentry box by the Depot gate. Down in the communications building, signals cypher and switchboard personnel awaited their 0300-hour change of shift. Up at the camp relieving communications staff were quietly dressing as their companions slumbered on.

At that same time, some 3,450 miles north of Lake Boga, on almost the same longitude, a Boeing B-29 Superfortress named *Enola Gay* was running up its four Wright engines, preparatory to take-off from Tinian, a tiny island nor nor east of Guam. On board this aircraft was an atomic bomb, *Little Boy*, the release of which some five hours 45 minutes later, over Hiroshima, would contribute significantly to the shortening of World War II. *Fat Boy*, a more powerful atomic bomb, was dropped on Nagasaki three days later. Japan would soon surrender!

In an early morning broadcast on Wednesday 15 August, Prime Minister Chifley formally announced Japan's surrender signalling the celebration of VP (Victory in the Pacific) Day.

At the Depot, the air raid siren whined non-stop as personnel tooted transport vehicle horns: Signal horns powered with batteries aboard beached Catalinas blasted away. Up at the camp a similar cacophony heralded the cessation of hostilities. The rattle of Pratt and Whitney motors added to the day's hullabaloo as Dutchman Lieutenant J Weyer first shot up the Depot then alighted in Y-87. Test pilot Flight Lieutenant Lance Steadman added to the VP Day activity with a 45-minute flip in A24-40.

At Swan Hill celebrations were no less spontaneous. In crowded Campbell Street people hugged and cheered. The *Guardian* later proclaimed:

> Joyful Crowds Throng Streets as Welcome News Spreads … With bugle and with drum, with pipe playing and voices raised in joy, Swan Hill literally "went to town" … The news swept the trading centre and shops closed when bells and sirens spread the historic message.

Celebrating crowds joined both impromptu and official street processions with the Swan Hill Municipal Brass Band leading the afternoon's march. No. 1 FBRD personnel looked impressive as they joined with WWI veterans, Scouts, Guides and members of the National Fitness League proceeding to a service of thanksgiving at the Cenotaph and Town Hall. Mayor Harold Harrison and the Reverend Angus Eadie presided at the service which was followed by community singing. That evening "fun and frolic" continued in the streets "until the early hours of the morning".

All that day, demands on Swan Hill telephone exchange facilities were acute as operator Nell Frazer recalled:

> I was on the night shift when the war ended. When I came on at 11pm that night the streets were still full of people celebrating. We relieved the late shift who were anxious to join the excitement. They said they had not been able to hold the switch board all day and left us with a stack of calls to get off and clear. With all the noise outside it was the only night we ever had to put on headsets at that late hour.

While the war was won, many Depot personnel still had months of productive service ahead of them. The SWPA remained alive with aircraft, and RAAF Catalinas of Nos. 111, 112, 113 and 115 ASR Flights responded to a multitude of distress signals. Within weeks mercy missions were flown to Singapore, Manila, Borneo and other Pacific areas repatriating Australian POWs whose health was sufficiently sound to endure the hours of airborne travel.

As they had in war, Catalinas proved themselves versatile in peace, even to being used as a meeting platform during preliminary Japanese surrender negotiations. Mariners too, were in their element flying a multitude of movements requiring relocation of personnel, materials and equipment.

Another type of waterborne aircraft entered Lake Boga skies on 17 August with the arrival of three Vought Sikorsky OS2U Kingfisher floatplanes. A48-5, A48-13 and A48-15 arrived from St George's Basin, near Jervis Bay. In deference to RAAF regulations in not allowing Kingfishers to fly long distances overland, the seaplanes had arrived at Boga via the southeast coast having alighted at Malacoota and Williamstown. A typical flying time was seven hours and 35 minutes.

At this date, the RAAF still had nine additional Kingfisher seaplanes on charge, all of which arrived for storage at Boga over the following fourteen days. Upon arrival, ground personnel waded out, attaching pre-delivered beaching gear after which the aircraft were slipped. The Kingfishers were then tied down in open storage beside the No. 1 slipway, their engines suppressed, then covered with tarpaulins as were the "glasshouse" cockpits. Several Depot personnel, including electrician LAC Ross Hepworth, recalled one Kingfisher pilot causing great consternation when executing a low pass. He flew straight through the power lines, cutting electricity to the area.

The end of August saw the loss of a most valued Catalina, a loss keenly felt by those who had crewed her. On 30 August at Darwin, the RAAF's first acquired Catalina A24-1 was damaged beyond repair during an aborted take-off towards the shoreline. This occurrence was most unfortunate as Catalinas A24-1 and A24-2 were to have been present in Singapore during the formal Japanese surrender.

Throughout the full 44 months of the war against Japan, RAAF flying boat crews had flown continual offensive, surveillance and support operations. A total of 229 Catalina crew members were dead. With the war won, Depot strength stood at 868 personnel.

The first full month of peace began with the arrival of the last two Catalinas to be taken on charge by the RAAF. Still bearing RAF registration JX634 from its diverted UK order, this was the first beached and would become A24-386. Allowing for the two breaks in Catalina registration numberings, the RAAF had now received a total of 168 Catalinas.

By now numerous postings to discharge Units were received, and No. 1 FBRD officers mess closed on 12 September. However, normal business continued during the month albeit at a reduced rate: only fifteen flying boats were serviced during the four and a half months to the end of the year. The Dutch Catalina Y-45 arrived carrying two additional crews who would fly out the newly serviced Y-87 and Y-57.

No. 107 Squadron's Kingfisher A48-11 arrives at Lake Boga in August 1945.

A view of No. 1 FBRD in October 1945 after a welcome rain. From the 15 August cessation of hostilities, aircraft buildup at Lake Boga increased as flying boats and seaplanes arrived for storage. Seven flying boats are moored offshore while several can be seen within the Depot, both in hangars and in the open. The stored Kingfisher floatplanes are in the centre of the image. (Frank Smith)

Meanwhile on a flight from Morotai to Perth while repatriating POWs, Flight Lieutenant Lindsay Oats had occasion to land in the Swan River, Perth, where his Catalina became stranded on a sand bank. Oats recalled:

> The flight was one of many carried out after VJ Day. On the way home we made a night flight from Darwin to Perth. Having studied the maps provided showing all the various hazards, I succeeded in setting the aircraft down on one of the very sand banks I was supposed to carefully avoid. Thinking we must have been on the outer edge of a bank I gunned the engines, only to find that by so gunning we went further on to the sand bank. After some ribald comments from the crew, we were taken ashore to enjoy Perth's hospitality while ground crews successfully towed the aircraft free.

Catalina A24-303 had departed Boga on 18 September to join three flying boats on a mercy mission to the Philippines, repatriating Australian POWs. Squadron Leader Ron Foskett headed the four Cat flight and by 29 September the RAAF flying boats swung at their moorings off Sangley Point, near Manila.

Now departing Manila, the Australian POWs made an early 0545 hour start on their homeward journey. As they climbed from a USN crash boat into their waiting RAAF aircraft, the men were each issued with Red Cross kits, medication and a pullover for altitude flying. Each Catalina carried twenty eager yet anxious Australians. Squadron Leader Ron Foskett:

> The first thing I noticed on my Aircraft Manifest was the weight of my passengers, they had obviously been "fattened up" since the end of hostilities. We flew eight hours fifteen minutes from Manila to Morotai, and then on 30 September nine hours fifteen minutes to Darwin. This was one of the big emotional moments of my life, when we began faintly to discern the low-lying shores of Bathurst and Melville Islands, then Darwin. The men came up to the cockpit one by one and saw Australia again. There weren't too many dry eyes. We flew sixteen hours fifteen minutes to Brisbane on 1 October and on to Rose Bay the following day.

Eighteen air movements were recorded at Boga in October including the last opportunity to witness a Mariner take-off when A70-11 departed for Rose Bay.

Depot personnel welcomed their fifth new CO as Squadron Leader George F De La Rue assumed command of No. 1 FBRD on 13 November. Arriving on a posting from No. 1 Personnel Depot, the new CO had joined the RAAF in 1935 and been appointed to a commission in February 1942. By month's end Depot strength had depleted by 127 personnel since August's cessation of hostilities.

During December eleven Catalinas and a solitary Mariner (see below) flew into Boga. The reduction of personnel continued but by Christmas Day cooks cursed with the arrival of a convoy of 51 RAAF vehicles with 96 personnel of No. 4 Repair and Salvage Unit, *en route* to Tocumwal. The unit departed on Boxing Day.

At 1512 hours on Christmas Day, Mariner A70-10 touched down and proceeded to buoy up on the tossing lake. The Mariner crew had recently been headlined in the *Brisbane Telegraph* after a forced landing off Queensland. The Mariner had recently left Port Moresby with a crew of five plus twenty-five Australian service personnel bound for Bowen, Brisbane and points south. During the flight the starboard engine failed necessitating a forced landing off Woodgate near Bundaberg.

The pilot Flight Lieutenant McKnight had eased the Mariner down in a nose up attitude and successfully executed a full stall alighting. "RESCUERS REACH FLYING BOAT" the *Brisbane Telegraph* front page read dramatically the following day. The aircraft had been forced to wallow its way through the long night, at times taxiing from the coast with its port engine. With few windows and no horizon upon which to focus, many on board suffered acute sea sickness. The following morning RAAF sea rescue launch 0-32 took the great flying boat in tow and by 1400 hours had the Mariner securely moored in midstream off the Bundaberg wharf.

With Mariner A70-10 now safely at Boga, McKnight proceeded to Swan Hill where he later partook of an evening meal with a family whose daughter had earlier prevailed upon Duty Signalman LAC Bernie Fitzgerald to bend RAAF regulations and radio an inflight invitation to the Mariner captain.

Nineteen Catalinas had arrived at Boga during December while five departed. A pattern was now emerging whereby many aircraft were flying in for storage with no orders for further maintenance. The RAAF flying boat squadrons that had served Australia so valiantly were now being disbanded or were in receipt of signals advising their days were numbered. November had seen No. 42 Squadron disbanded while Nos. 20 and 43 Squadrons were posted south to Rathmines before termination. After more than six years dedicated service, No. 11 Squadron was disbanded on 20 December.

On New Year's Eve, Depot strength comprised 22 Officers, 595 airmen and 50 WAAAFs, for a total of 667 personnel. For three and a half years, Lake Boga had lived through a boom-town existence. In the years ahead that heightened experience would return to normal.

Mariner A70-10 after arrival at Lake Boga on Christmas Day 1945 for storage.

CHAPTER 5
1946-1947

The coming year would see the dramatic rationalisation of No. 1 FBRD as the facility was wound down. The first contingent of WAAAFs had arrived at Lake Boga in November 1942: now the last would depart. Clad in their khaki summer drabs the remaining 50 WAAAFs vacated the camp in January. Their CO Squadron Leader De La Rue now inspected the silent quarters that had for 38 months resounded to the voices of so many dedicated and happy personnel. With the WAAAFs so suddenly gone, the absence of female companionship was keenly felt in the airmen's workplace and social environment.

A pattern was now emerging as flying boats were progressively withdrawn from service. While the number of waterborne aircraft was building at Lake Boga, many Catalina amphibians and late model Boeing-built flying boats were still operational or arriving for storage at Rathmines. However, Lake Boga RAAF air movements would continue until October 1947, albeit infrequently.

January's arrivals out-numbered departures five to one with nineteen Catalinas and one Mariner alighting. Flying boats were now towed ashore joining other old warriors, their great wings outstretched as they jostled for space like giant migratory birds in an overcrowded rookery.

In RAAF Operation Record Books around Australia, 1946 was a year in which increasing numbers of personnel received orders advising "completion of duties", "ceased attachment", "termination of appointment" or "departed on posting for discharge".

With a clutter of craft stored throughout the Depot or moored-up on the lake the perimeter fence was extended. Barbed wire was run from the Depot towards the Boga township parallel to the highway enclosing the DCA property. Many Catalinas were now towed into this newly secured area, lined up in rows, wing tip to wing tip facing the Murray Valley Highway. With canvas covered engines, most of them rested, out in all weathers dusty and forlorn until their disposal in 1947.

January had not been without its dust storms. When one storm abated, it was found to have taken an unusual toll, littering gardens and streets with dead galahs. A total of 164 personnel departed on posting during the month, leaving a strength of 503 personnel, 17 of whom were officers.

Five flying boats arrived during the first two days of February. On its final flight, Martin Mariner A70-12 alighted captained by Flight Lieutenant Phil Mathieson. Mathieson was one of those few captains who had flown the RAAF's one, two, three and four engined flying boats: Walrus and Sikorsky, Catalina and Mariner, Dorniers and the great Sunderlands. Of these, the Flight Lieutenant's preference was the Sunderland for its weight and handling characteristics.

The Dutch now paid a last call. On 19 February Catalina Y-69 made a refuelling stop on its flight from Perth to Rose Bay. With Lieutenant Hoebink's departure for Sydney at 1330 hours, the Depot's link with the flying Dutchmen of the Royal Dutch Naval Air Service was severed. The link had commenced in the darker days of March 1943.

Martin Mariner A70-11, beached for the final time in February 1946. (Bill Muller)

February 25 saw the last multi-aircraft movement to Lake Boga with the arrival of Mariner A70-11 and Catalinas A24-21, -62, -65 and -57.

In mid-February, the Rathmines based No. 11 Squadron was disbanded. Nos. 43 and 20 Squadrons were disbanded in March. By now parties of RAAF officers and government officials were observed inspecting the Depot buildings with a view to their disposal. February saw a further reduction in Depot strength of 198 personnel.

On 1 March No.1 Flying Boat Repair Depot was redesignated as "Care and Maintenance Unit, RAAF, Lake Boga". In what proved to be the final departure of a Lake Boga serviced Catalina, Flight Lieutenant Shepherd took-off in mid-March aboard A24-381 for Rathmines.

With its great tail rising 32 feet 10 inches above its keel, No. 40 Squadron's Sunderland A26-4 at Lake Boga on 15 April 1946. (Michael Smith)

If Lake Boga residents held any doubts as to the RAAF unit's future, the sight of two heavily laden convoys of Depot motor vehicles departing for Melbourne on 22 March provided a further indication of the drastic rationalisation. Gone were the flat top tenders that had so regularly transported personnel to and from their workplace and on day leave to Swan Hill. The fuel tanker, light tenders, tip truck, Dodge utility and the two Indian motorcycles were also all gone.

Depot personnel pose on a Catalina mainplane. The Catalina, engines shrouded, rests on stilts near the magazine building.

March also saw the arrival of Mariners A70-4 and -7, while Squadron Leader John O'Donnell took over from Squadron Leader De La Rue as commanding officer. The last day of the month saw the final exodus of Depot vehicles take place and was accompanied by the closure of the Sick Quarters at Castle Donnington. The small number of sick had been transferred to the Swan Hill District Hospital.

Twelve months had passed since Lake Boga had seen its first Sunderland flying boat. Now, on 12 April Wing Commander Vic Hodgkinson alighted in A26-4 on dead calm water. The aircraft departed 30 minutes later, returning on 15 April when flights were offered to investors in the Second Security Loan. There was no shortage of people eager to experience a flight in this enormous craft, its predecessors being synonymous with RAAF flying boat operations in Britain.

The Depot's Post Office "No. 428, RAAF, Australia" closed on 1 May 1946 with all mail from that date handled by the Lake Boga Postmaster. That same day, the last Martin Mariner alighted, A70-8. These rugged aircraft had done a sterling job in the service of No. 41 Squadron executing a multitude of tasks in wartime and the immediate post-war period. Once ashore, A70-8 joined the eleven companion Mariners sitting silently nosed up on their cumbersome beaching gear. While two of these bulky boats were hangared the balance languished in the open. They were great stilled monoliths, awaiting an ignominious fate.

Authority was granted by HQ Southern Area on 3 May, for the closure of all signals facilities. Valuable radio equipment was then crated and departed the area. By month's end the Depot's technical library was transferred to HQ Southern Area and in June the Depot Medical Officer departed.

By August the Commonwealth Disposals Commission began advertising the sale of Lake Boga Catalina aircraft, declared surplus. The sale, newspapers advised, would be by tender and would close on 6 September 1946. Prospective purchasers noted that seven Catalinas and one airframe were for sale, all suitable for conversion to civilian operation. The flying boats in question were among RAAF registrations A24-4 to A24-30.

Like giant birds in a rookery, RAAF flying boats rest in storage at No. 1 FBRD, circa 1946. (Brett Freeman)

Meanwhile working parties arrived to transfer machine tools to Richmond, NSW, while stores were relocated to Rathmines. The Lake Boga unit, whose strength a year earlier had been 868, now numbered just twelve airmen and five officers. The adjacent parade ground was now in use as the new Lake Boga football oval.

In September, 72 flying boats remained at Boga comprising 47 Catalinas, twelve Martin Mariners, twelve Kingfishers and one Supermarine Walrus. Of eight Catalinas sold by tender, Kingsford Smith Aviation purchased seven and WR Carpenter Pty Ltd bought A24-26, then sold it to Port Melbourne boat builders Botterill and Fraser.

Boga's CO Squadron Leader John O'Donnell relinquished command on 1 October 1946, departing to Tocumwal on his new posting. Assuming command, was the Depot's seventh and final CO, Squadron Leader W Symons who arrived from Wagga.

The new year of 1947 saw sufficient work remaining on hand for the tiny Depot staff as high summer ushered in further dust and discomfort. All of the Depot's 46 buildings were intact, still housing various levels of RAAF stores, equipment and spare parts, while quantities of documents stood stored in the administration buildings.

During March the Commonwealth Disposals Commission advertised the sale by tender of 36 stored Catalinas with the provision that those tendering must submit a price for the purchase of all 36 aircraft. By June, Sikorsky Kingfisher aircraft A48-9 and A48-13 were transported by road to Rathmines where they underwent restoration for allotment to the newly formed Australian Antarctic Flight. Three road transports departed Boga in July for Rathmines laden with crated Catalina parts.

With a view to later resale, the State Material Procurement Directorate purchased the accommodation camp's 70 odd buildings as a whole. The final day of July saw the few remaining RAAF personnel vacate the camp, relocating to the Depot. In recording progress at this date the last sentence of CO, Squadron Leader Symon's July report should not be taken too literally:

> The camp area (complete) was purchased by the State Material Procurement Directorate and all personnel were transferred to the Depot area. This served a dual purpose. Personnel were now able to sleep on the job, and it enabled the easy policing of the Depot.

A decision by the State Materials Procurement Directorate to auction camp buildings in October produced enquiries from Government instrumentalities, industrial firms and civilians. When the Department of Aircraft Production advertised the sale of remaining Depot equipment by tender, the need to escort prospective purchasers on inspection caused considerable disruption to stock take and crating tasks.

On 6 August the penultimate visit to Lake Boga of an RAAF Catalina occurred when A24-381 alighted carrying a ferry crew who would, nine days later, depart in the sole remaining Supermarine Walrus, HD874, for Rathmines. This aircraft was also allotted to the Australian Antarctic Flight.

In September, Boga was drenched with long awaited rains which forced visiting vehicles to confine their movements to "hard standing areas only". Parties from Parkes and Kerang now inspected the hangars as RAAF flat top transports departed for Rathmines with three additional Kingfisher floatplanes.

As no tender was accepted as a whole for the purchase of the Boga based Catalinas, 34 Catalinas and beaching gear were now advertised in metropolitan newspapers for sale separately. With tenders closing on 19 September, aircraft inspection was brisk. Among prospective buyers was wartime visitor and ex-RAAF Cat captain Stewart Middlemiss who purchased A24-58 on behalf of the newly registered Barrier Reef Airways. Together with his partner Poulson, Middlemiss had also acquired four other Catalinas from Rathmines.

Harry Purvis, a pioneer aviator, now arrived on the scene. At this date Purvis was director of the Fairfax Newspaper Group's Sydney Morning Herald Flying Services, and selected Catalina A24-59. This aircraft was additional to three others acquired from Rathmines.

On Sunday morning 9 October, an RAAF Catalina visited Lake Boga for the final time. Yachtsmen watched as Squadron Leader Robin Gray DFC alighted and then taxied towards the Depot. After loading, with difficulty, a quantity of Kingfisher spare parts for delivery to Rathmines, Catalina A24-385 taxied clear of the yachtsmen and at 1515 hours, commenced its full throttled run across the lake, the foaming wake cut short as the hull broke free. As it reached the distant shoreline, the Catalina banked slowly, a diminishing silhouette. Minutes more and its motors had faded to silence.

A month later, on 8 November a civilian Cat, VH-BDP (formerly A24-26), returned to Lake Boga, its crew no doubt in search of spare parts. Two days later this flying boat departed and was the last take-off recorded during the RAAF's reign.

On 12 November 1947, the Care and Maintenance Unit RAAF, Lake Boga, was disbanded when FC Brown was installed as caretaker. Several weeks later on 17 December the first auction of the camp land took place when seven lots were offered, subject to the owners of buildings still on the site being given 60 days' notice.

Aircraft numbers stored awaiting collection or destruction comprised 36 Catalinas, twelve Martin Mariners and six Kingfisher floatplanes. The buildings remained, subsequent to sale and removal. The unit's strength numbered just a single officer and five other ranks.

In over five years of Depot life, with more than 1,050 flying boat arrivals and departures and an estimated 800 test flights, plus associated aerial stunting, no aircraft met with mishap. This circumstance must surely place No. 1 FBRD in a unique status among RAAF annals.

The final page of the RAAF Unit's History Sheet concludes:

> During the life of the station, it turned out a large volume of work under very exacting conditions. Lake Boga itself is an ideal stretch of water for Catalinas, and even the largest flying boats could land without fear of trouble. To those who were stationed at No. 1 FBRD in its heyday, the station would look very drear and desolate should they pay a return visit. And so another unit which played a large part in the 1939-45 war passes.

Sgd W Symons S/Ldr, CO, CMU

12/11/47

Index of Names

Abbott, Bud 31
Allan, S/Ldr George U "Scotty" 11, 39, 41, 42, 45, 47, 49, 50, 52, 54, 57, 59-61
Allen, Judy 27
Allen, Leo 32
Ansell, Marj 16
Athom, Bill 14
Athom, Nell 14
Baker, Jim 22
Bannister, P/O 13
Barwick, Nana 27
Bate, Chaplain E 29
Beggs, Sister E 39
Benham, F/Sgt Laurie 25
Bickley, LAC Ellis 21
Bills, Annis 19
Bills, George 19
Bond, F/Lt "Chestie" 50
Bond, F/O Daniel 36
Brain, Lester 9, 10
Brameyer, Plt Sgt JD 50
Brammel, Sgt 17
Brown, W/O 63
Brown, FC 76
Browne, Dr FE "Dick" 19
Buchholz, Cpl Charles J 44
Bullman, F/Lt 66
Burrage, S/Ldr Reg 26, 36
Byrd, Lt (jg) Edwin "Red" 53-55
Chapman, F/Lt F 17
Chapman, S/Ldr 24
Cheverton, Lt 40
Chifley, Prime Minister 69
Clapson, Cpl Richard 15, 16, 32, 64
Claterboss, Lt 69
Clempson, Ivan 53
Corbert, AB 10

Corrie, F/Lt 66
Costello, Lou 31
Cramer (nee Thompson), Kath 40
Cumming, Elsie 43
Dally, Ma 10
de Bruyn, Cmdr AJ 40
De Castella, F/Lt 60
De La Rue, S/Ldr George F 72-74
De Lisle, Gordon 60
Delahunty, F/O Bert 26
Delaney, Laurie 32
Dohnt, LAC Norm 42
Don (nee Sands), Moyna 43
Doody, ACW Kathleen 57
Douglas, ASO 43
Dowling, Ted 32
Duigan, Gwynne 27
Duigan, F/Lt Terry 27, 32
Duncan, RB 12
Dunstan (nee Woosman), Ethel 28, 50
Dunstan, Stan 28
Eadie, Rev Angus 69
Edward, F/Lt Rodger 45
Essary, Lt (jg) Melvin 54
Etienne, F/Lt Armand 69
Fader, F/O Norm 21
Finkelstein, Joe 21
Fitzgerald, Alan 37
Fitzgerald, LAC Bernie 72
Fitzpatrick, James 23
Fleming, F/O 31
Foord, W/Cmdr Roy H 63-65
Ford, Dick 16
Foskett, S/Ldr Ron 51, 71
Franks, Ensign Herbert 47
Frazer, Charlotte "Nell" 19, 55, 70
Frost, CW 65

Fysh, Hudson 9, 10
Garden, W 21
Gates, Lt (jg) Robert 48
Goddard, LAC Albert 45
Graham, Lt 60
Gray, S/Ldr Robin 76
Greets, 2nd Lt Arjen Johan 49
Gryzen, Harry 40
Hampshire, S/Ldr John 29
Harrison, F/Lt 51
Harrison, Mayor Harold 69
Hartley, Lt (jg) Walter 38, 42
Hartvig, Lt (jg) Donald 38, 39
Hasselo, Lt Henk 57
Haydon, F/O Clem 25
Heggen, AVM Alan 60
Hepworth, LAC Ross 70
Herring, Audrey B 24
Higgins, Brian "Tubby" 27
Hill, F/Lt Jack 53
Hircock, LAC Edward Morris 53
Hirst, F/Lt Bob 22
Hodgkinson, W/Cdr Vic
Hoebink, Lt 73
Holland, Percy 12
Hollebon, LAC Ernest WM 33
Honan, P/O Bob 23, 24, 37, 51
Hornell, David 67
Hull, WO Cliff 45
Irvine, Roy 9
Jedeloo, Lt 37
Jenvey, LAC Jim 67
Johnson, F/Lt F 14
Jones, S/Ldr 26
Jones, AVM George 12, 32, 33
Jones, Gwen 43
Jones, Sgt Ken 36, 40

Kelso, Sect Off Norma HJM 57, 58
Kenney, Maj Gen George C 23
Kenzor (nee McCallum), ACW Palma 43, 57
Kingsford Smith, Sir Charles 11, 53
Knight, Sgt Bert 50
Knott, Capt Richard C 38
Knox, G 11
Lahodney, Lt William Jr 42
Laird, Bill 53
Le Bache, Dorrie 19
Leeflung, Lt 57
Leslie, F/Lt 31
Lowenstern, Cyril 10, 13
MacArthur, General 12, 57, 58
MacGregor, P/O D 14
Mainprize, F/Lt Max 66
Marr, S/Ldr Reg 51
Marsh, P/O P 17
Marshall, Alan 13, 14, 16, 17
Marshall, W/Cmdr Geoffrey Douglas 25, 33, 35, 36, 39-41
Marshall, Joyce 37
Marshall, Olive 14, 17
Marshall, Susan 37
Mason, F/Lt George 68
Mathieson, F/Lt Phil 73
Maychek, Lt 40
McKeand, F/Lt Lee 68
McKnight, F/Lt Noel 57, 72
McMahon, W/Cmdr John 9, 67
Meldrum, Beth 43
Merrit, L/Cmdr James Jr 63
Michels, Vladimir 10, 13
Middlemiss, Stewart 76
Miedecke, S/Ldr 67
Millard, Padre 50
Miller, F/Lt Bill 29, 3
Miller, Damien 52
Mills, Ollie 67
Mills, F/Lt Wally 65, 66
Moffat, S/Ldr George Stewart 14, 22, 25
Monk, LAC Ron 26
Monkton, S/Ldr Bryan 47, 60, 61

Moody, Sgt Tom 54
Morris, Charles 39
Moulton, LAC Clifford 58
Moxham, AC1 Norman 45
Murphy, A/Cde AW 65
Murphy, A/Cde J 31
Myers, F/O Gordon 14, 23, 41
O'Donnell, S/Ldr John "Jack" 22, 74, 75
O'Halloran, Matt 35
Oats, F/Lt Lindsay 71
Oldmeadow, S/Ldr Courtney 65
Owens, Bob 42
Oxen, W/Cmdr 32
Page, Sir Earle 9
Paragreen, Mary (nee Jones) 39
Parsons, Keith 49
Patrick, ASO Sheila 24, 42, 43
Petzke, Fred 9
Poole, G/Cpt 23
Potter, Sgt 36
Poulson 76
Pratt, S/Ldr 28
Purvis, Harry 76
Radford, S/Ldr J 45
Ranney, Lt Howard 40, 41
Rice, W/Cmdr Raymond S 60, 63
Riddell, F/O Jack 17, 18, 25
Robbie, Pere 26
Roberts, F/O K 68
Roker, Sam 31, 43
Rowe, F/Lt EA 14
Rowland (nee Pfeiffer), ACW Ethel 37
Rundle, Padre 50
Rundle, F/Sgt Lindsay 24
Ryan, Mrs 27
Ryan, Doc 27
Rynders, Lt 57
Sandow, Cpl Sandy 31
Schoech, Lt Cmdr 23
Scown, Charlie 22
Scown, Eddie 22
Seddon, P/O 13
Settle, Cpl Jack 18, 52, 63
Seymour, F/Lt Mike 17

Shanks, LAC Bill 11
Sharp, Capt RS 33
Shepherd, F/Lt 74
Sillers, Lt (jg) Colin 54, 56
Simoni, Lt 45
Steadman, F/Lt Lance 67, 69
Stevens, W/Cmdr 37
Stevens, Capt Paul 40
Stewart, F/Sgt Colin 14, 32
Stewart, Hughey 26
Stilling, S/Ldr Gordon 21, 23
Stokes, S/Ldr Tom 26
Symons, S/Ldr W 75, 76
Taylor, Cpl Bill 35
Tennant, S/Ldr John "Jock" 39, 41, 64
Toohey, S/Ldr J 33
Tozer, F/Sgt Allan David 69
Tweedie, LAC Doug 16
Udy, F/O Dick 51
Vernon S/Ldr David 31
Vincent, David 27, 53
Wackett, A/Cde EC 19, 29
Wackett, Lawrence J 19
Wallis, Jack D 9, 10
Wearne, S/Ldr Athol 59, 63
Webster, LAC Ranal "Ron" 31, 42, 50
Wendell, Charlie 23
Wentjes, Lt 61, 65
Weyer, Lt J 69
White, F/O "Bimbo" 23, 24
Whitlock, Bill 40
Whitlock, Marjorie 40
Wilkins, June 24
Williams, Gwen 63
Wingrave, P/O W 14
Womersley, Frank 25
Womersley, Jack 16
Wood, Betty 42
Wood, Peggy 42
Wood, S/Ldr Sam 44, 47
Worner, Neil 13, 26, 30, 36
Wright, F/O Dudley 52
Zeillo, Carmelia 16
Zierk, Keith 29

www.ingramcontent.com/pod-product-compliance
Ingram Content Group UK Ltd.
Pitfield, Milton Keynes, MK11 3LW, UK
UKHW061203180426